Student Study Guide

to accompany

Hole's Human Anatomy and Physiology

Eighth Edition

David Shier
Jackie Butler
Ricki Lewis

Prepared by
Nancy A. Sickles Corbett
Camden County Visiting Nurse Association, Inc.
Cherry Hill, NJ

WCB McGraw-Hill

Boston Burr Ridge, IL Dubuque, IA Madison, WI New York San Francisco St. Louis
Bangkok Bogotá Caracas Lisbon London Madrid
Mexico City Milan New Delhi Seoul Singapore Sydney Taipei Toronto

WCB/McGraw-Hill

*A Division of The **McGraw-Hill** Companies*

Student Study Guideto accompany or for use with
HOLE'S HUMAN ANATOMY AND PHYSIOLOGY

3 4 5 6 7 8 9 0 QPD/QPD 9 0 9

ISBN 0-697-34206-9

www.mhhe.com

CONTENTS

TO THE STUDENT

A study guide attempts to do as the term implies: that is, to guide your study so that your learning efforts are most efficient.

This study guide is based on several beliefs: (1) Learning occurs best when the learner is active rather than passive; (2) learning is easiest when the material is organized in simple units: and (3) the learner can best evaluate what he or she knows well, is unsure of, or does not know.

The stsudy guide chapters correspond to the chapters in the text, *Hole's Human Anatomy and Physiology*, eighth edition, by David Shier, Jackie Butler, and Ricki Lewis. The elements of the study guide chapters and their purposes are described below.

1. *Overview*. The learning objectives at the beginning of each chapter in the text are arranged in groups according to broad, general concepts presented in the chapter. The overview also contains a purpose statement that offers a rationale for studying the chapter.
2. *Chapter Objectives*. The chapter objectives from the text are listed.
3. *Focus Question*. The focus question helps you focus your study of each chapter.
4. *Mastery Test*. The mastery test, taken before reading the chapter, is designed to help you identify the following:
    a. The concepts you already know.
    b. Those concepts you need to clarify.
    c. Those concepts you do not know.

If you are using a study guide for the first time, you may be unfamiliar with this type of testing. It is important for you to realize that this test is *for your information*. Its purpose is to help you learn where to concentrate your learning efforts; therefore, it is best not to guess at any answers.

5. *Study Activities*. A variety of study activities helps facilitate the study of the principal ideas of each chapter.

The study activities should be done after you have read the chapter carefully, concentrating on those areas that the mastery test indicates you do not know.

The first activity in each chapter is a vocabulary exercise, concentrating on words or word parts appropriate to each chapter. You are asked to define these as you understand them and then compare your definitions with those in the chapter. You may find it helpful to define terms orally and in writing. If you have a tape recorder, you may use it as a study device.

After the vocabulary exercise, you may be asked to describe a process, label a diagram, fill in a chart, or observe the function of a body part in yourself or in a partner. (This partner may be a classmate or a cooperative family member.) These are written activities, but you may also find it helpful to repeat them orally.

After you complete the study activities, retake the mastery test. A comparison of the two scores will indicate the progress you have made. You may also wish to set a learning goal for yourself, such as a score of 70%, 80%, or 90%, on the mastery test after completing your study of a chapter. If you have not attained your goal, the mastery test results can show where you need additional study.

The answers to the mastery test are at the end of the study guide. You can compare your responses to the review activities by referring to the appropriate page numbers in the text. Each major section in the study guide is identified by a Roman numeral and the title of the corresponding section in the text. The activities in the study guide are lettered, and the corresponding pages in the text are noted after the activity.

You are responsible for your own learning. No teacher can assume that responsibility. A study guide can help you direct your study more efficiently, but only you can control how well and how completely you use the guide.

CHAPTER 1
INTRODUCTION TO HUMAN ANATOMY AND PHYSIOLOGY

OVERVIEW

This chapter begins the study of anatomy and physiology by defining the disciplines (objective 1), and explaining the characteristics and needs that are common to all living things (objectives 2 and 3). It introduces a basic mechanism necessary to maintain life (objectives 4 and 5), as well as the relationship of increasingly complex levels of organization in humans (objective 6). The study of levels of organization continues with the identification of body cavities and the organs found within each cavity (objectives 7 and 8). The membranes associated with the abdominopelvic and thoracic cavities are described (objective 9). The functions of the various organ systems as well as the organs associated with each system are described (objectives 10 and 11). Finally, the language used to describe relative positions of body parts, body sections, and body regions is presented (objective 12).

This chapter defines the characteristics and needs common to all living things and the manner in which the human body is organized to accomplish life processes. The language peculiar to anatomy and physiology is also introduced.

CHAPTER OBJECTIVES

After you have studied this chapter, you should be able to:

1. Define *anatomy* and *physiology,* and explain how they are related.
2. List and describe the major characteristics of life.
3. List and describe the major requirements of organisms.
4. Define *homeostasis* and explain its importance to survival.
5. Describe a homeostatic mechanism.
6. Explain the levels of organization of the human body.
7. Describe the locations of the major body cavities.
8. List the organs located in each major body cavity.
9. Name the membranes associated with the thoracic and abdominopelvic cavities.
10. Name the major organ systems and list the organs associated with each.
11. Describe the general functions of each organ system.
12. Properly use the terms that describe relative positions, body sections, and body regions.

FOCUS QUESTION

How is the human body organized to accomplish those tasks that are essential to maintain life?

MASTERY TEST

Now take the mastery test. Do not guess. Some questions may have more than one correct answer. As soon as you complete the test, correct it. Note your successes and failures so that you can read the chapter to meet your learning needs.

1. Study of the human body began with earliest humans because .
   a. our early ancestors were curious about the world around them.
   b. they were as interested in their body parts and their functions as we are today.
   c. of their concern with illness and injury.

2. Which of the following factors set the stage for the development of modern science?

a. a belief that spirits or gods controlled sickness and health

b. the growing experience of medicine men as they treated the sick with herbs and potions

c. the belief that natural processes were caused by forces that could be understood

d. the ability to ask questions and record the answers

3. What two languages form the basis for the language of anatomy and physiology?

4. The branch of science that studies the structure (morphology) of body parts is _______________.

5. The branch of science that studies what body parts do and how they do it is _______________.

6. The function of a part is (always/sometimes/never) related to its structure.

7. List those characteristics that are common to all living organisms.

a.

b.

c.

d.

e.

f.

g.

h.

i.

j.

8. The physical and chemical changes or reactions that occur in the body are called ____________.

9. The vital signs include

a. temperature.

b. heart rate.

c. respiratory rate.

d. reflex activity.

10. The most abundant chemical substance in the human body is _______________.

11. Food is used as an _______________ source, to build new _______________ _______________, and to participate in chemical reactions.

12. Oxygen is used to release _______________.

13. An increase in temperature (increases/decreases) the rate of chemical reactions.

14. The application of force on an object or substance is _______________.

15. Atmospheric pressure plays a part in _______________.

16. Homeostasis means

a. maintenance of a stable internal environment.

b. integrating the functions of the various organ systems.

c. preventing any change in the organism.

17. Body temperature is maintained around a set point of 37° C.

a. True

b. False

18. The set point for the body's temperature is controlled by the

a. skin.

b. circulatory system.

c. lungs.

d. hypothalamus.

19. Blood sugar (is/is not) maintained by a negative feedback mechanism.

20. Positive feedback mechanisms usually lead to (health/illness).

21. The smallest particle in the human body is the

a. molecule.

b. atom.

c. cell.

Questions 22–25. Match the structures listed in the first column with the functions listed in the second column.

Structure	**Function**
_____ 22. atoms, molecules, macromolecules	a. groups of cells that have a common function
_____ 23. cells	b. chemical structures required for life
_____ 24. tissues	c. allow life to continue despite changing environments and reproduce to continue the species
_____ 25. organisms	d. simplest living units

26. List the five levels of organization of the body in order of increasing complexity, beginning with the cell.

27. The portion of the body that contains the head, neck, and trunk is called the _______________ portion.

28. The arms and legs are called the _______________ portion.

29. The two major cavities of the axial portion of the body are the _______________ cavity and the _______________ cavity.

30. The inferior boundary of the thoracic cavity is the _______________.

31. The heart, esophagus, trachea, and thymus gland are located in the _______________ of the thoracic cavity.

32. The pelvic cavity is

a. the lower one-third of the abdominopelvic cavity.

b. the portion of the abdomen that contains the reproductive organs.

c. the portion of the abdomen surrounded by the bones of the pelvis.

33. The visceral and parietal pleural membranes secrete a serous fluid into a potential space called the _______________ _______________.

34. The heart is covered by the _______________ membranes.

35. The peritoneal membranes are located in the _______________ cavity.

36. Match the systems listed in the first column with the functions listed in the second column.

___ 1. nervous system	a. reproduction
___ 2. muscular system	b. processing and transporting
___ 3. cardiovascular system	c. integration and coordination
___ 4. respiratory system	d. support and movement
___ 5. skeletal system	
___ 6. digestive system	
___ 7. lymphatic system	
___ 8. endocrine system	
___ 9. urinary system	
___ 10. reproductive system	

37. Which of the following positions of body parts is/are in *anatomic* position?

a. palms of hands turned toward sides of body

b. standing erect

c. arms at side

d. face toward left shoulder

38. Terms of relative position are used to describe
   a. the relationship of siblings within a family.
   b. the importance of the various functions of organ systems in maintaining life.
   c. the location of one body part with respect to another.

39. A sagittal section divides the body into
   a. superior and inferior portions.
   b. right and left portions.
   c. anterior and posterior portions.

40. The terms *epigastric, hypochondriac,* and *iliac* are examples of _______________ _______________.

41. Ultrasonography involves the use of _______________; MRI creates an image of body parts using a _______________ field.

42. Which of the following body parts will be well visualized using ultrasonography?
   a. lungs
   b. heart
   c. bone
   d. uterus

43. MRI (is/is not) a good technique to use to study soft tissue such as the brain.

Study Activities

I. Definition of Key Terms

Define the following terms used in this chapter.

anatomy

appendicular

axial

cardiovascular

digestion

excretion

homeostasis

metabolism

negative feedback

organelle

organism

pericardial

peritoneal

physiology

pleural

reproduction

respiration

thoracic

visceral

II. Introduction and Clinical Example (pp. 2–3)

A. While examining Judith, what question might the nurse ask her in order to focus her examination?

B. Why are the vital signs such important indicators of an individual's physiologic state?

C. What knowledge directed health care personnel's assessment and diagnosis of Judith's problems? Explain your answer.

D. Why did the study of the human body begin with attempts to understand illness and injury rather than with attempts to understand the human body?

III. Anatomy and Physiology (pp. 3–4)

Explain how the structure of the following parts is related to the function given.

fingers : grasping

heart : pumping

blood vessels : moving blood in the proper direction

mouth : receiving food

teeth : breaking food into smaller pieces

tongue and cheeks : mixing food particles with saliva

IV. Characteristics of Life (pp. 4–5)

A. Describe the following characteristics of life.

movement

responsiveness

reproduction

growth

respiration

digestion

absorption

assimilation

circulation

excretion

B. What is metabolism?

C. Why are observations of the vital signs important to nurses and physicians?

V. Maintenance of Life (pp. 5–8)

A. Match the terms in the first column with the statements in the second column that define their role in the maintenance of life.

____ 1. water	a. essential for metabolic processes
____ 2. food	b. governs the rate of chemical reactions
____ 3. oxygen	c. creates a pressing or compressing action
____ 4. heat	d. necessary for release of energy
____ 5. pressure	e. provides chemicals for building new living matter

B. Homeostasis

1. Define *homeostasis.*

2. How is body temperature maintained at 37°C (98.6°F)?

3. Where is the "set point" for body temperature controlled?

4. Describe negative and positive feedback mechanisms. Give examples of each.

VI. Levels of Organization (pp. 8–9)

Arrange the following structures in increasing levels of complexity: atoms, organ systems, organelles, organism, organs, macromolecules, cells, tissue, molecules.

VII. Organization of the Human Body (pp. 10–18)

A. The dorsal cavity is subdivided into the _____________ cavity and the _____________ cavity.

B. Answer the following concerning the ventral cavity.

1. The ventral cavity is subdivided into the ______________ cavity and the ______________ cavity.
2. The ______________ divides the ventral cavity.
3. List the viscera found in each portion of the ventral cavity.

C. List the four smaller cavities of the body.

D. Fill in the blanks.
1. The walls of the thoracic cavity are lined with a ______________ membrane called the ______________ ______________.
2. The lungs are covered by the ______________ ______________.
3. Why is the pleural cavity called a potential space?

E. Name and describe the membranes covering the heart.

F. The linings of the abdominopelvic cavity are the ______________ ______________ and the ______________ ______________.

G. Fill in the following chart.

Structure and Function of Organ Systems

Function	Organ System	Organs in System
Support and movement	1.	1.
	2.	2.
Integration and coordination	1.	1.
	2.	2.
Processing and transporting	1.	1.
	2.	2.
	3.	3.
	4.	4.
	5.	5.
Reproduction: female	1.	1.
Reproduction: male	2.	2.

VIII. Anatomical Terminology (pp. 18–24)

A. Use the illustration on this page to specify the terms that describe the relationship of one point on the body to another.

1. Point (*a*) in relation to point (*d*).

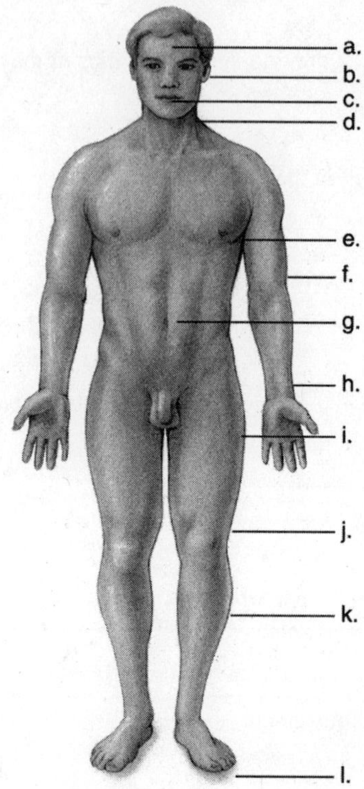

2. Point (*f*) in relation to point (*h*).

3. Point (*g*) in relation to point (*i*).

4. Point (*l*) in relation to point (*j*).

5. Point (*i*) in relation to point (*g*).

6. Point (*c*) in relation to point (*a*).

B. Use the illustration on this page to perform the following exercises.

1. Draw a line through the drawing to indicate a midsagittal section. How is this different from a frontal section?
2. Draw a line through the drawing to indicate a transverse section.
3. Define cross section, longitudinal section, and oblique section.

4. Locate and label the following body regions: epigastric, umbilical, hypogastric, hypochondriac, lumbar, and iliac. Locate these regions on yourself or on a partner.

5. Use lines *a l* to locate and label the following body parts on the diagram: antebrachium, antecubital, axillary, brachial, buccal, cervical, groin, inguinal, mammary, ophthalmic, palmar, pectoral.

IX. Ultrasonography and Magnetic Resonance Imaging (p. 10–11)

A. Describe ultrasonography.

B. Describe magnetic resonance imaging.

X. Clinical Focus Question

List the ways that health professionals use their knowledge of anatomy and physiology to assess and evaluate their clients' health.

When you have completed the study activities to your satisfaction, retake the mastery test and compare your performance with your initial attempt. If there are still areas you do not understand, repeat the appropriate study activities.

CHAPTER 2
CHEMICAL BASIS OF LIFE

OVERVIEW

This chapter introduces some basic concepts of chemistry, a science that studies the composition of substances and the changes that occur as basic elements combine. It explains how substances combine to make up matter (objectives 1–5), how substances are classified as acid or base (objective 6), and the organic and inorganic substances that make up the living cell (objectives 7 and 8).

Knowledge of basic chemical concepts will enhance your understanding of the functions of cells and of the human body.

CHAPTER OBJECTIVES

After you have studied this chapter, you should be able to:

1. Explain how the study of living material depends on the study of chemistry.
2. Describe the relationships between matter, atoms, and molecules.
3. Discuss how atomic structure determines how atoms interact.
4. Explain how molecular and structural formulas are used to symbolize the composition of compounds.
5. Describe three types of chemical reactions.
6. Define pH.
7. List the major groups of inorganic substances that are common in cells.
8. Describe the functions of various types of organic substances in cells

FOCUS QUESTION

How is chemistry related to the structure and function of living things and their parts?

MASTERY TEST

Now take the mastery test. Do not guess. Some questions may have more than one correct answer. As soon as you complete the test, correct it. Note your successes and failures so that you can read the chapter to meet your learning needs.

Questions 1–5 Match the structures listed in the first column with the functions listed in the second column.

Structure	**Function**
____ 1. atom	a. molecular building blocks and energy sources for living cells
____ 2. molecule	b. the operating instructions for living cells—the genes
____ 3. electrolyte	c. smallest complete unit of an element
____ 4. carbohydrates, lipids, protein	d. two or more atoms joined together
____ 5. nucleic acids	e. molecule that gives rise to ions (charged particles) in the internal environment

6. The symptoms of Slone's disease are due to
   a. an excess of trypsin.
   b. an excess of copper.
   c. an excess of serotonin.
   d. lack of tryptophan.
7. The discipline that deals with the chemistry of living things is called ______________.
8. What is matter? In what forms can it be found?

9. The basic units of matter are _______________.

10. When two elements are found in combination, the substance is called a _______________.

11. Which of the following substances is *not* an element?

a. iron
b. bronze
c. oxygen
d. hydrogen

12. Carbon, hydrogen, oxygen, and nitrogen are examples of _______________ elements.

13. Many trace elements are important parts of _______________.

14. What elements are most plentiful in the composition of the human body?

15. An atom is made up of

a. a nucleus.
b. protons.
c. neutrons.
d. electrons.
e. all of the above

16. Match the following:

____ 1. neutron	a. positive electrical charge
____ 2. proton	b. negative electrical charge
____ 3. electron	c. no electrical charge

17. The atomic number of an element is determined by the number of _______________.

18. The atomic weight of an element is determined by adding the number of _______________ and the number of _______________.

19. An isotope has the same atomic _______________ but different atomic _______________.

20. The atoms of the same element have the same number of _______________, but may vary in the number of _______________.

21. When an isotope decomposes and gives off energy, it is

a. unstable.
b. radioactive.
c. explosive.

22. The interaction of atoms is determined primarily by the number of _______________ they possess.

23. Atomic radiation that travels the most rapidly and is the most penetrating is

a. alpha.
b. beta.
c. gamma.

24. The time it takes for one-half of the amount of an isotope to decay to a nonradioactive form is its _______________.

25. Radioactive isotopes of iodine can be used both to diagnose and treat thyroid problems because

a. iodine concentrates in the thyroid gland and only the thyroid gland.
b. iodine is intrinsically destructive to the thyroid gland.
c. Both
d. Neither

26. An element is chemically inactive if

a. it has a high atomic weight.
b. its outer electron shell is filled.
c. it has an odd number of protons.

27. An ion is

a. an atom that is electrically charged.
b. an atom that has gained an electron.
c. an atom that has lost an electron.
d. all of the above

28. An electrovalent bond is created by

a. a positive ion and a negative ion attracting each other.
b. two or more positive ions combining.
c. two or more negative ions combining.

29. In forming a covalent bond, electrons are

a. shared by two atoms.
b. given up by the atom.
c. taken up by the atom.
d. none of the above

30. The ionizing radiation to which people in the U.S. are exposed is generated by

a. radioactive substances that occur naturally in the earth's surface.
b. medical and dental X rays.
c. the use of radioactive materials for power generation.
d. the inclusion of radioactive substances in consumer products.

31. A molecule is made up of two or more atoms of _______________ _______________ _______________.

32. A compound is formed when atoms of _______________ elements combine.

33. $C_6H_{12}O_6$ is an example of a _______________ formula.

34. 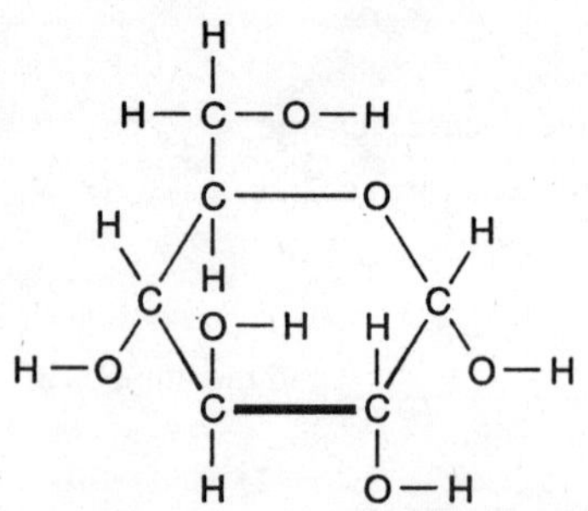 is an example of a _______________ formula.

35. Two major types of chemical reactions in which molecules are formed or broken are called _______________ and _______________.

36. A + B → is a _______________.

37. The speed of a chemical reaction is affected by the presence or absence of a _______________.

38. Match the following.

____ 1. accepts hydroxide ions in water — a. acid
____ 2. product of an acid and a base reacting together — b. base
____ 3. releases hydrogen ions in water — c. salt

39. What is the pH of a neutral solution? _______________ An acid solution? _______________
A basic solution? _______________

40. A blood pH of 7.8 indicates a condition known as _______________.

41. Organic substances are most likely to dissolve in _______________ or _______________.

42. Which of the following substances does *not* play a role in the polarization of the cell membrane?

a. sodium ions
b. potassium ions
c. bicarbonate ions
d. sulfate ions

43. Identify the following cell constituents with an *O* if they are organic and an *I* if they are inorganic: a. water ( ), b. carbohydrate ( ), c. glucose ( ), d. oxygen ( ), e. protein ( ), f. carbon dioxide ( ), g. fats ( ).

44. Carbohydrates contain atoms of _______________, _______________, and _______________.

45. The building blocks of fat molecules are _______________ _______________, and _______________.

46. The building blocks of protein molecules are _______________ _______________.

47. The function of nucleic acids is to

a. store information and control life processes.
b. act as receptors for hydrogen ions.
c. neutralize bases within the cell.

48. A technique that allows visualization of body structures in three dimensions is _______________ _______________.

Study Activities

I. Definition of Key Terms

Define the following terms used in this chapter.

atom

carbohydrate

compound

decomposition

electrolyte

element

inorganic

ion

isotope

lipid

molecule

nucleic acid

organic

protein

synthesis

II. Introduction (p. 37)

A. Describe the biochemical error that causes Slone's disease.

B. What is the subject matter that chemists study?

C. What subdivision of chemistry is of particular interest to physiologists? Why?

III. Structure of Matter (pp. 37–46)

A. Answer the following concerning elements and atoms.

1. Anything that has weight and takes up space is _______________.
2. Basic substances are called _______________.
3. Tiny, invisible particles that make up basic substances are called _______________.
    a. What is a bulk element? Give examples.

b. What is a trace element? Give an example.

c. What is an ultratrace element? Give an example.

4. Two or more particles of the same basic substance form a _______________.
5. Two or more particles of different basic substances form a _______________.

B. Fill in the following chart.

Element	Symbol	Element	Symbol
Oxygen		Sodium	
Carbon		Magnesium	
	H	Cobalt	
Nitrogen			Cu
	Ca		F
	P	Iodine	
	K		Fe
Sulfur			Mn
	Cl	Zinc	

C. Answer the questions that pertain to the diagram on this page.

What is the atomic number of this atom?

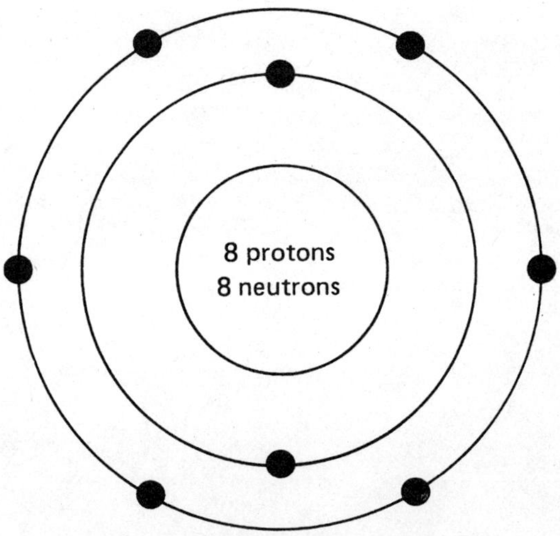

What is its atomic weight?

How many electrons are needed to fill its outer shell?

Is this element active or inert?

Identify this element.

D. Answer the following questions about isotopes.

How does the isotope of any element differ from the element?

How does this difference affect the way an isotope interacts with other elements? Explain your answer.

What is a radioactive isotope?

List the three types of radiation.

Describe what is meant by the term *half-life* of an isotope.

Describe at least two clinical uses of radioactive isotopes.

E. Draw diagrams of an electrovalent bond and a covalent bond. How are these bonds different?

F. Answer the following questions about ionizing radiation.

1. Define ionizing radiation.

2. Describe the effects of ionizing radiation on the health of living things.

3. List the sources of ionizing radiation. Which of these sources can be controlled by humans?

G. Answer the following questions that pertain to the accompanying diagram.

1. Give the molecular formula of this substance.

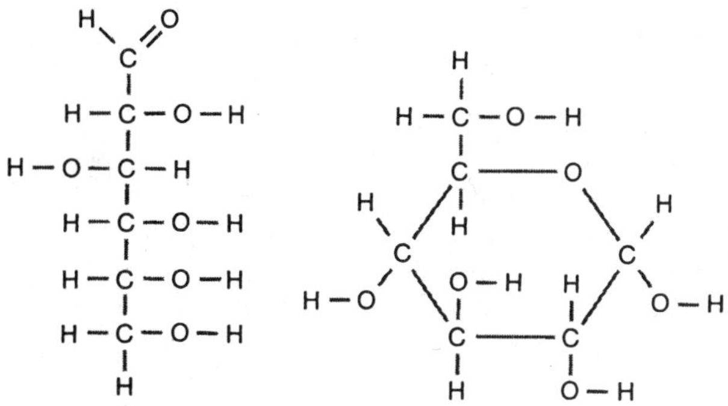

Identify this compound.

Is it an organic or an inorganic substance? Why?

H. 1. Identify the type of chemical reaction in these equations.

$AB \rightarrow A + B$ $\quad$ $A + B \rightarrow AB$ $\quad$ $A + B \rightleftharpoons AB$ $\quad$ $AB + CD \rightarrow AD + CD$

2. What is the role of a catalyst in a reversible reaction?

I. Identify these substances as being acid, base, or salt.

HCl	$Mg(OH)_2$	H_3PO_4
$NaOH$	$NaHCO_3$	$MgSO_4$
H_2SO_4	$NaCl$	$Al(OH)_3$

J. Answer the following concerning acid and base concentrations.

1. What is meant by pH?

2. Identify these substances as being acid or base.

carrot: pH 5.0	tomato: pH 4.2
milk of magnesia: pH 10.5	lemon: pH 2.3
human blood: pH 7.4	distilled water: pH 7.0

3. What is alkalosis and what are its clinical signs and symptoms?

4. What is acidosis and what are its clinical signs and symptoms?

IV. Chemical Constituents of Cells (pp.46–56)

A. What is the role of the following inorganic substances in the cell? Be specific.

water

oxygen

carbon dioxide

inorganic salts or electrolytes (Na, K, Ca, HCO_3, PO_4)

B. How do NO (nitric oxide) and CO (carbon monoxide) affect body function?

C. Answer the following concerning carbohydrates.

1. What is the role of carbohydrates in maintaining the cell?

2. What are the elements found in carbohydrates?

3. Describe simple and complex carbohydrates.

D. Answer the following concerning lipids.

1. What elements are found in lipids?

2. What is the role of lipids in maintaining the cell?

3. Substances classified as lipids are ______________, ______________, and ______________.
4. Fats are composed of ______________ and ______________.
5. Fats containing single carbon-carbon bonds are ______________.
6. Fats containing double-bonded carbon atoms are ______________.
7. Describe the composition and characteristics of a phospholipid molecule.

8. List commonly occurring steroids in body cells.

E. Answer the following concerning proteins.

1. What is the role of protein in maintaining the cell?
2. What elements are found in protein?
3. The building materials of a protein are ______________ ______________.
4. A protein molecule that has become disorganized and lost its shape is said to be ______________.
5. How is the shape of a protein molecule maintained?

F. Describe the roles of the two nucleic acids, DNA and RNA.

G. Describe PET imaging. What makes it especially useful?

H. Describe CT scanning. What makes it a particularly useful technique?

V. Clinical Focus Question

Vomiting and diarrhea lead to loss of both fluid and electrolytes. How might these losses be most effectively treated? Why is prompt treatment important?

When you have finished the study activities to your satisfaction, retake the mastery test and compare your results with your initial attempt. If you are not satisfied with your performance, repeat the appropriate study activities.

CHAPTER 3
CELLS

OVERVIEW

This chapter deals with the basic unit of structure of the human body—the cell. It explains the makeup of a composite cell and the contribution of each of the organelles to cellular function (objectives 1, 2, and 4). It introduces the various mechanisms used to transport material into and out of the cell (objectives 3 and 6). The process of cell reproduction and the factors that control cell reproduction (objectives, 5, 7, 8 and 9) are discussed.

An understanding of cell structure is basic to understanding how cells support life at the cellular and organismic levels.

CHAPTER OBJECTIVES

After you have studied this chapter, you should be able to:

1. Explain how cells differ from one another.
2. Describe the general characteristics of a composite cell.
3. Explain how the components of a cell membrane provide its functions.
4. Describe each kind of cytoplasmic organelle and explain its function.
5. Describe the cell nucleus and its parts.
6. Explain how substances move into and out of cells.
7. Describe the cell cycle.
8. Explain how a cell reproduces.
9. Describe several controls of cell reproduction.

FOCUS QUESTION

How does the structure of cellular organelles contribute to and support the functions of each organelle and, in turn, the cell?

MASTERY TEST

Now take the mastery test. Do not guess. Some questions may have more than one correct answer. As soon as you complete the test, correct it. Note your successes and failures so that you can read the chapter to meet your learning needs.

1. A small number of people do not contract AIDS when exposed to the HIV virus because
   a. their cell membrane contains a substance toxic to the HIV virus.
   b. the nuclear membranes in these individuals is impermeable to the HIV virus.
   c. of abnormal glycoproteins on the inner surface of the cell membrane.
   d. their cell membrane lacks a docking site critical to HIV penetration.

Questions 2–5. Match the structures listed in the first column with the functions in the second column.

Structure

____ 2. cells

____ 3. cell membrane

____ 4. nucleus

____ 5. various cytoplasmic organelles

Function

a. performance of various cellular activities such as protein synthesis, metabolism, and cellular reproduction

b. control of all cellular activities; contains the genetic material of the cell

c. controls movement of substances into and out of the cell; allows the cell to respond to certain stimuli

d. the smallest living units

6. Bacteria have a simpler cell structure than human cells and lack a nucleus. They are called

a. eukaryotic.
b. prokaryotic.
c. archaeal.
d. akaryotic.

7. The only cell in the human body that is visible without the use of a microscope is the __________.

8. Which of the following statements about the hypothetical composite cell is/are true?

a. It is necessary to construct a composite cell because cells vary so much based on their function.
b. It contains structures that occur in many kinds of cells.
c. It contains only structures that occur in all cells, although the characteristics of the structure may vary.
d. It is an actual cell type chosen because it occurs most commonly in the body.

9. The two major portions of the cell, each surrounded by a membrane, are the ______________ and the ______________.

10. The organelles are located in the

a. nucleolus.
b. cytoplasm.
c. cell matrix.
d. cell membrane.

11. Which of the following statements about the cell membrane is *false?*

a. An intact cell membrane is essential to the life of the cell.
b. The cell membrane is composed of phospholipids and protein.
c. The cell membrane does not participate in chemical reactions.
d. The cell membrane is selectively permeable.

12. The cell membrane is essential to the ability to receive and respond to messages.

a. True
b. False

13. What characteristic of the cell membrane makes it impermeable to such substances as water, amino acids, and sugars?

a. an intercellular matrix that makes it difficult for water-soluble substances to get close to the cell membrane
b. the phosphate groups that form the outermost and innermost layers of the cell membrane
c. the fatty acid portions of phospholipids that make up the middle layer of the cell membrane
d. the fibrous proteins that span the width of the cell membrane

14. Which of the following substances on the cell membrane surface helps cells to recognize and bind to each other as well as recognizing "non-self" substances such as bacteria?

a. proteins
b. cholesterol
c. glycerol
d. glycoproteins

15. Water-soluble substances such as ions cross the cell membrane via

a. active transport mechanisms.
b. protein channels.
c. carrier mechanisms.
d. phagocytosis.

16. Receptors on the cell membrane are composed of

a. carbohydrates.
b. proteins.
c. fats.
d. triglycerides.

17. Faulty ion channels can cause disease and sudden death.

a. True
b. False

18. A class of drugs that are used to treat hypertension and angina affect which of the following transmembrane channels?

a. calcium channels
b. sodium channels
c. chloride channels
d. potassium channels

19. The intercellular junction characterized by tight fusion between cells to form sheetlike layers of cells is called a

a. tight junction.
b. desmosome.
c. gap junction.

20. Cell adhesion molecules are proteins that

a. provide the "glue" for cells to adhere to each other permanently.

b. attract cells to an area in which they are needed.

c. are part of a mechanism that allows cells to interact in a totally different manner than usual.

21. The blood-brain barrier shields brain cells from toxins by

a. stimulation of brain cells to secrete a special lipid to coat the cell membranes.

b. modifications of the walls of capillaries in the brain.

c. increasing the protein in the cell membrane of capillaries and brain cells.

d. creating a fatty matrix in which brain cells are embedded.

22. The organelle that functions as a communication system for the cytoplasm is the ______________ ______________.

23. The chemical activity in the endoplasmic reticulum results in

a. synthesis of protein.

b. dissemination of amino acids.

c. oxidation of glucose.

d. synthesis of lipid molecules.

24. Which of the following statements about ribosomes is *false?*

a. Ribosomes are part of rough endoplasmic reticulum.

b. Ribosomes are made up of protein and RNA.

c. Ribosomes secrete proteins used as enzymes.

d. The secretion of ribosomes is utilized only within the cell.

25. The secretion of the Golgi apparatus is ______________.

26. The mitochondria are also called the ______________ of cells.

27. Cells with very high energy requirements are likely to have (more/less) mitochondria than cells with low energy requirements.

28. Which of the following organelles is most likely to be of interest to evolutionary biologists?

a. nucleolus

b. cell membrane

c. lysosome

d. mitochondria

29. The enzymes of the lysosome function to

a. control cell reproduction.

b. digest bacteria and damaged cell parts.

c. release energy from its storage place within the cell.

d. control the Krebs cycle.

30. Peroxisomes are found most commonly in the cells of the

a. liver.

b. heart muscle.

c. kidney.

d. central nervous system.

31. Which of the following statements about the centrosome is/are *true?*

a. It is located near the nucleus.

b. The centrioles of the centrosome function in reproduction.

c. The centrosome is concerned with the distribution of genetic material.

d. all of the above

32. Mobile, hairlike projections that extend outward from the surface of the cell are called ______________ or ______________.

33. A membranous sac formed when the cell membrane folds inward and pinches off is a

a. microtubule.

b. cytoplasmic inclusion.

c. vesicle.

d. lysosome.

34. Thin, threadlike structures found within the cytoplasm of the cell are called ______________ and ______________.

35. The structures that float in the nucleoplasm of the nucleus are the ______________ and the ______________.

36. Which of the following organelles may show evidence of shrinking or clumping as individuals age?

a. nucleus

b. lysosomes

c. cell membrane

d. chromosomes

37. The process that allows the movement of gases and ions from areas of higher concentration to areas of lower concentration until equilibrium has been achieved is called _______________.

38. The process by which substances are removed through the cell membrane by a carrier molecule is called _______________ _______________.

39. The process by which water moves across a semipermeable membrane from areas of low concentration of solute to areas of higher concentration is called _______________.

40. A hypertonic solution is one that

a. contains a greater concentration of solute than the cell.

b. contains the same concentration of solute as the cell.

c. contains a lesser concentration of solute than the cell.

41. The process by which molecules are forced through a membrane by hydrostatic pressure that is greater on one side than on the other is called _______________.

42. The process that uses energy to move molecules or ions across a concentration gradient from an area of lower concentration to an area of higher concentration is called _______________ _______________.

43. The process by which cells engulf liquid molecules by creating a vesicle is called _______________.

44. A process that allows cells to take in molecules of solids by surrounding them to create a vesicle is called _______________.

45. Which of the following statements best describes what happens when solid material is taken into a vacuole?

a. A ribosome enters the vacuole and uses the amino acids in the "invader" to form new protein.

b. A lysosome combines with a vacuole and digests the enclosed solid material.

c. The vacuole remains separated from the cytoplasm and the solid material persists unchanged.

d. Oxygen enters the vacuole and "burns" the enclosed solid material.

46. The process by which specific kinds of particles are moved through the cell membrane even when a substance is present only in very small concentrations is called _______________ _______________ _______________.

47. The selective and rapid transport of a substance from one end of a cell to the other is known as

a. endocytosis.

b. transcytosis.

c. pinocytosis.

d. exocytosis.

48. The process that ensures duplication of DNA molecules during cell reproduction is _______________.

49. Match these events of cell reproduction with the correct description.

____ 1. Microtubules shorten and pull chromosomes toward centrioles.	a. prophase
____ 2. Chromatin forms chromosomes; nuclear membrane and nucleolus disappear.	b. metaphase
____ 3. Chromosomes elongate, and nuclear membranes form around each chromosome set.	c. anaphase
____ 4. Chromosomes become arranged midway between centrioles; duplicate parts of chromosomes become separated.	d. telophase

50. The period of cell growth and duplication of cell parts is called _______________.

51. The process by which cells develop unique characteristics in structure and function is called _______________.

52. The process of cloning involves

a. transfer of the nucleus of a differentiated cell to replace the nucleus of a fertilized egg.

b. fertilization of an egg outside the body.

c. destroying the nucleus of a fertilized egg.

d. transferring a fertilized egg from the body of one individual to another.

53. Identify the following cells using *A* if they reproduce continually throughout life, *B* if they reproduce when an injury occurs, and *C* if they do not reproduce.

a. skin
b. liver
c. blood-forming cells
d. nerve cells
e. intestinal lining

54. The theory that a cell increases in volume more than it can increase its surface area attempts to explain

a. cell reproduction.
b. cell excretion.
c. cell metabolism.
d. none of the above

55. Hyperplasia and anaplasia are characteristics of _______________.

Study Activities

I. Definition of Key Terms.

Define the following terms used in this chapter.

active transport

centrosome

chromosome

cytoplasm

cytoskeleton

differentiation

diffusion

endocytosis

endoplasmic reticulum

equilibrium

exocytosis

extracellular

facilitated diffusion

filtration

Golgi apparatus

intercellular

intracellular

lysosome

micrometer

mitochondrion

mitosis

nucleolus

nucleus

osmosis

permeable

phagocytosis

pinocytosis

ribosome

transcytosis

vesicle

II. Introduction (p. 60)

A. Describe the cell membrane anomaly that protects some rare individuals from contracting the HIV virus despite exposure to the virus.

B. Cell sizes are measured in units called _______________.

C. What human cell is visible to the unaided eye?

D. How do cells differ from one another?

III. A Composite Cell (pp. 60–78)

A. Answer the following concerning a composite cell.

1. List the three major components of a cell.

2. Subdivisions of cytoplasm that serve specific functions necessary for cell survival are called _______________. Examples of these are _______________.
3. The _______________ directs the overall activities of the cell.
4. Cells of the human body are eukaryotic. What does this mean? Are bacteria also eukaryotic?

B. Answer the following concerning the cell membrane.

1. Describe the general characteristics and functions of the cell membrane.

2. Answer the following questions about the structure of the cell membrane.

   a. Describe the chemical structure of the cell membrane. How is this structure related to its function?

   b. Label the "heads" and "tails" of the phospholipid molecules in the following diagram of a cell membrane.

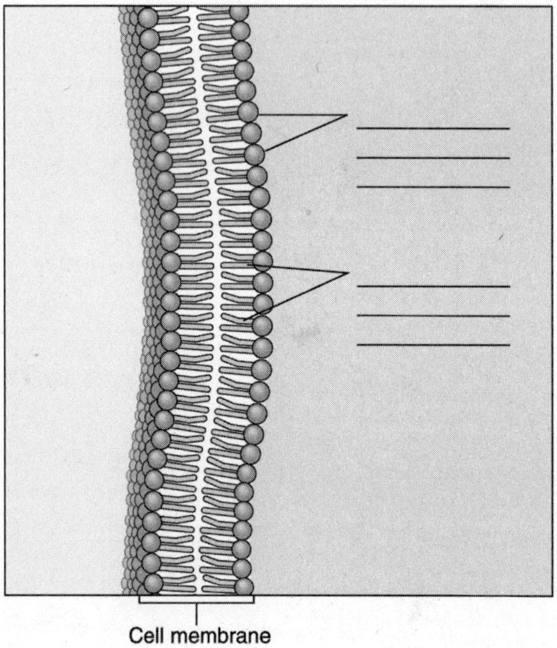

   c. Draw a fibrous protein and a globular protein in the preceding cell membrane diagram.

   d. What are the functions of these proteins?

   e. Draw carbohydrate groups on the surface of the membrane. Identify whether these molecules are found on the outer surface of the membrane or on the inner surface of the membrane.

   f. What are the functions of these glycoprotein groups?

3. Explain why some substances can pass through the cell membrane while others cannot.

4. Describe how the substances that cannot pass through the cell membrane enter the cell.

5. Describe the problems created by faulty ion channels.

6. Describe the intercellular junction of cells that form sheetlike layers.

7. An intercellular junction that serves as a "spot weld" is called a _______________. Where is this kind of intercellular junction found?

8. An intercellular junction that joins cells together in the form of tubular channels that allow exchange of ions is found in the ______________.

9. How do cell adhesion molecules function to support interaction between cells?

10. Describe the brain-blood barrier.

C. Describe the appearance of cellular cytoplasm.

D. Fill in the following chart.

Structure and function of cellular organelles

Organelle	Structure	Function
Endoplasmic reticulum		
Ribosome		
Golgi apparatus		
Mitochondria		
Lysosomes		
Peroxisomes		
Centrosome		
Cilia and flagella		
Vesicles		
Microfilaments and microtubules		
Nuclear envelope		
Nucleus		
Nucleolus		
Chromatin		

E. How do abnormalities of peroxisomal enzymes affect health? Give examples.

F. Describe the structure and function of the cell nucleus.

IV. Movements Into and Out Of the Cell (pp. 78–87)

A. List the substances that enter and leave the cell through the cell membrane.

B. List the physical processes used in the movement of substances across the cell membrane.

C. List the physiological processes used to move substances across the cell membrane.

D. Answer the following concerning movement through cell membranes.

1. In what direction do the molecules of solute move in diffusion?

2. When does diffusion stop?

3. What substances in the human body are transported by diffusion?

E. Describe facilitated diffusion. How is it different from diffusion?

F. What transport mechanism is illustrated in the following pictures? Describe the mechanism.

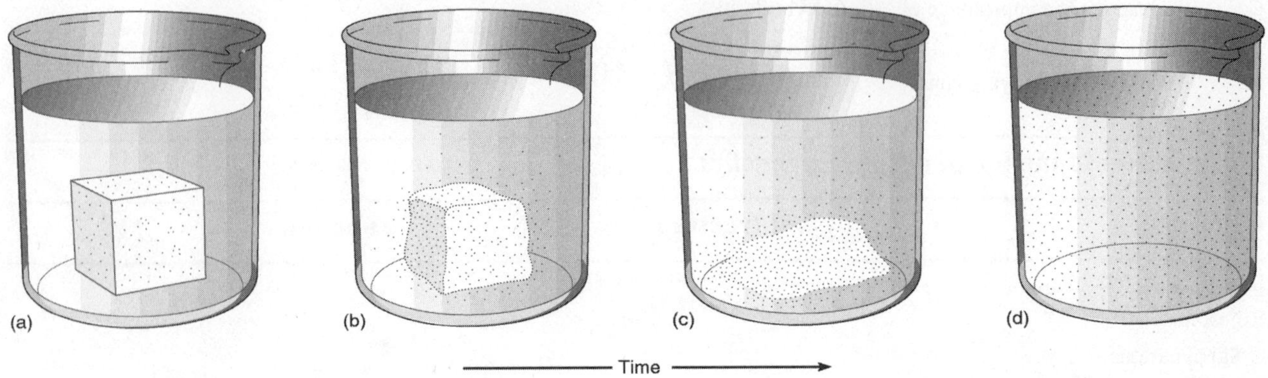

G. Following are three micrographs of a red blood cell in solutions of varying tonicity. Label the solution in each drawing and explain what is happening and why.

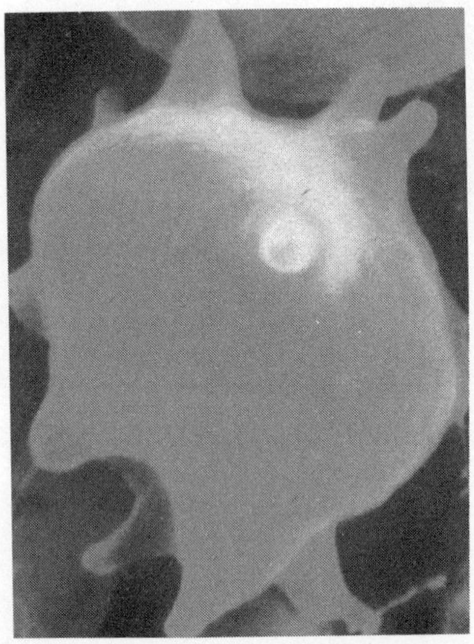

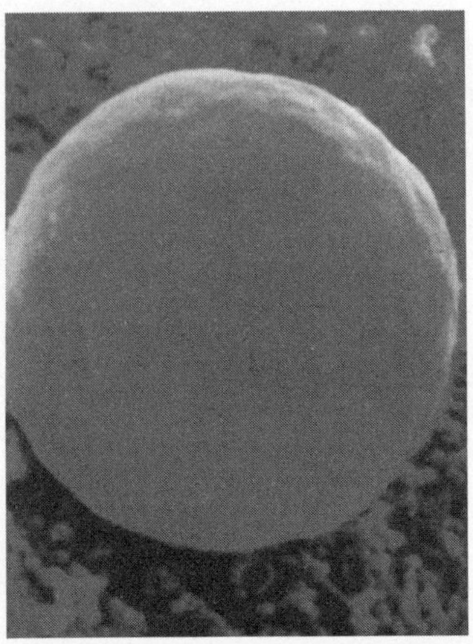

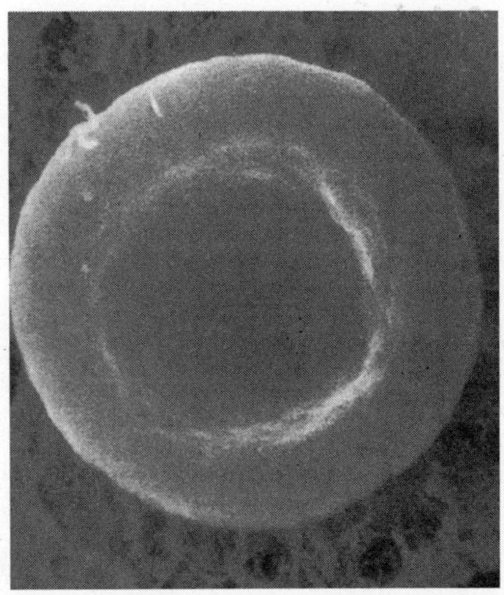

H. Answer these questions concerning the illustration at right.

1. What process is illustrated here?

2. What provides the force needed to pull the liquid through the solids?

3. Where within the body does this process occur?

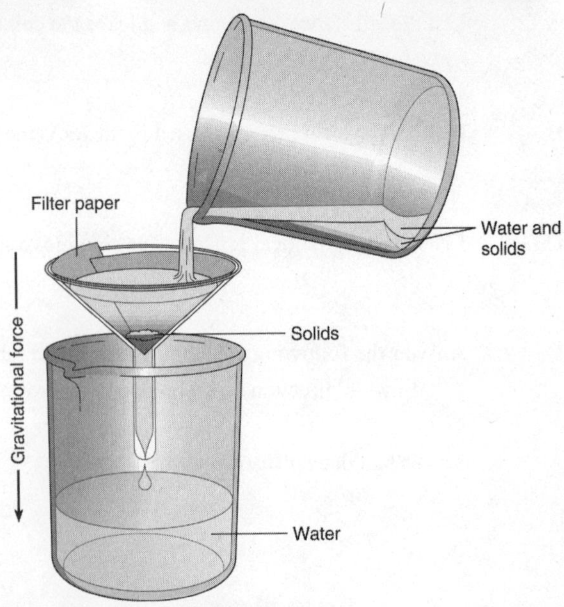

I. Answer these following concerning the accompanying diagram.

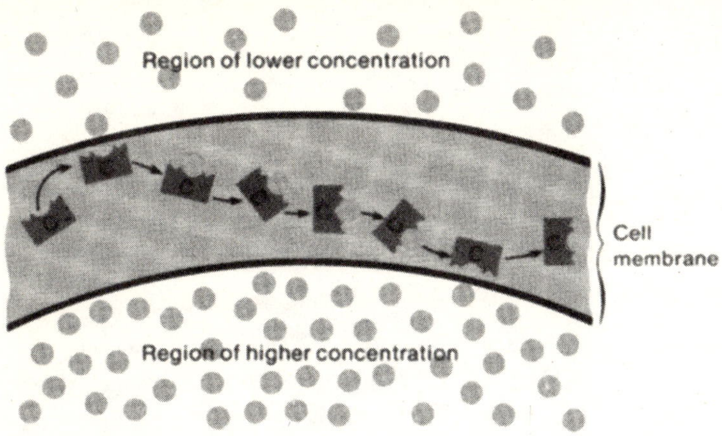

1. What transport mechanism is illustrated here?

2. What provides the necessary force for this process?

3. What is the source of this force?

4. How are molecules transported across the cell membrane?

5. What substances are transported by this mechanism?

J. Answer those questions concerning the accompanying diagram.

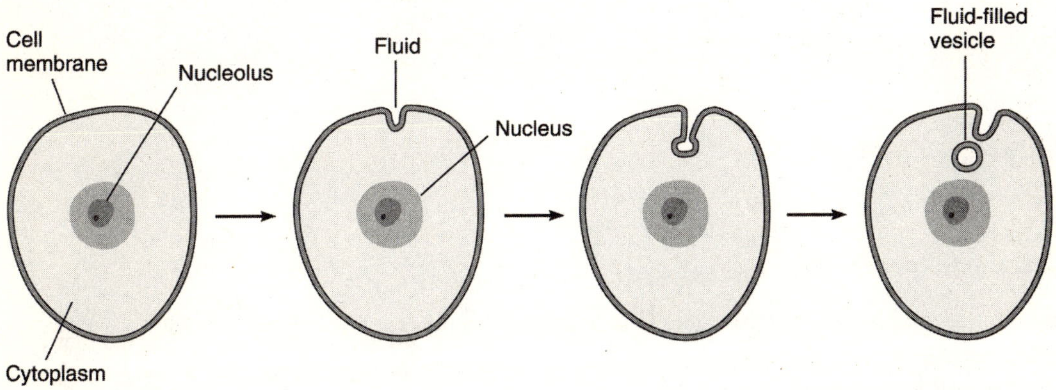

1. What transport mechanism is illustrated in this drawing?

2. Describe how this mechanism works and explain what substances are transported across the cell membrane.

3. How important is this mechanism to cell survival?

4. Describe the pathologic mechanism of the "storage diseases."

K. Answer the questions concerning the following diagram.

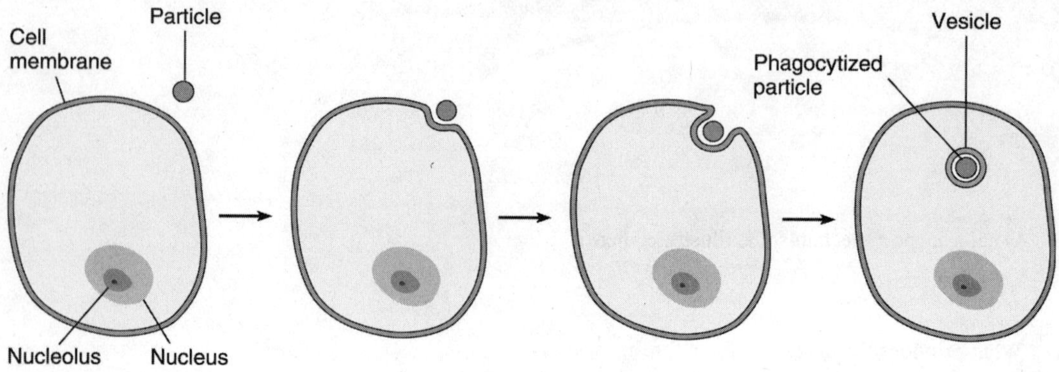

1. What mechanism is illustrated in this drawing?

2. What organelles are involved?

3. What kinds of foreign objects are transported?

4. How is this mechanism important to cell survival?

5. Describe receptor-mediated endocytosis.

6. Describe transcytosis.

V. The Cell Cycle (pp. 88–93)

A. The series of changes that a cell undergoes from its formation until its reproduction is called its ______________ ______________.

B. What two processes are involved in cell reproduction?

C. Describe each of the following events in mitosis.

prophase

metaphase

anaphase

telophase

D. Describe the events of cytoplasmic division.

E. The process by which cells develop differences in structure and function is _______________ _______________.

F. Answer the following questions about the process of cloning.

1. Describe the process used by researchers to clone an adult sheep.

2. What are the potential benefits and harms that could be derived from cloning human beings?

VI. Control of Cell Reproduction (pp. 93–97)

A. Describe two mechanisms that are thought to control reproduction of cells.

B. Fill in the following chart.

Type of cell	Differences in cell reproduction
Cells that reproduce continually	
Cells that reproduce in response to injury	
Cells that do not reproduce	

C. List the characteristics that make cancer cells different from normal cells.

D. Describe the clinical uses of epidermal growth factor (EGF).

VII. Clinical Focus Question

How is knowledge of the mechanism of cellular transport and cellular reproduction applied to the treatment of disease?

When you have completed the study activities to your satisfaction, retake the mastery test and compare your performance with your initial attempt. If there are still areas you do not understand, repeat the appropriate study activities.

CHAPTER 4
CELLULAR METABOLISM

OVERVIEW

This chapter deals with two basic cellular processes: the utilization of energy and the use of genetic information to control cell processes. Specifically, this chapter discusses how enzymes control cell metabolism, how energy is released and made available to the cell, and how carbohydrates, lipids, and proteins are metabolized (objectives 1–6). It also explains how genetic information is stored, how it is used to control cell processes, and how the cell changes when genetic information is altered (objectives 7–10).

An understanding of life processes at the cellular level is basic to understanding how life processes occur at more complex levels—as in the whole organism.

CHAPTER OBJECTIVES

After you have studied this chapter, you should be able to:

1. Distinguish between anabolism and catabolism.
2. Explain how enzymes control metabolic processes.
3. Explain how the actions of cellular respiration release chemical energy.
4. Describe how energy is made available for cellular activities.
5. Describe the general metabolic pathways of carbohydrate, lipid, and protein metabolism.
6. Explain how metabolic pathways are regulated.
7. Describe how DNA molecules store genetic information.
8. Explain how protein synthesis relies on genetic information.
9. Describe how DNA molecules are replicated.
10. Explain how genetic information can be altered and how such a change may affect an organism.

FOCUS QUESTION

How does eating a diet high in carbohydrates help long-distance runners performances?

MASTERY TEST

Now take the mastery test. Do not guess. Some questions may have more than one correct answer. As soon as you complete the test, correct it. Note your successes and failures so that you can read the chapter to meet your learning needs.

1. Bacterial toxins cause disease and death by damaging the
   a. cell membrane.
   b. mitochondria.
   c. nuclear envelope.
   d. Golgi apparatus.
2. The metabolic process that synthesizes materials for cellular growth and uses energy is called _______________.
3. The metabolic process that breaks down complex molecules into simpler ones and releases energy is called _______________.
4. The process by which two molecules are joined together to form a more complex molecule and water is called
   a. dehydration synthesis.
   b. chemical bonding.
   c. atomization.

5. Glycerol and fatty acids become joined to form water and

a. cholesterol.
b. lard.
c. fat molecules.
d. wax.

6. Amino acids are joined by a peptide bond to form water and _______________.

7. The process by which water is added to a complex molecule to break it down as represented by the equation $C_{12}H_{22}O_{11} + H_2O \rightarrow C_6H_{12}O_6 + C_6H_{12}O_6$ is called _______________.

8. Anabolism must be (greater than/equal to/less than) catabolism to maintain the life of the cell.

9. The form of energy used to activate metabolic reactions is

a. electrical.
b. heat.
c. mechanical.
d. light.

10. Enzymes are composed of

a. lipids.
b. carbohydrates.
c. proteins.
d. inorganic salts.

11. Enzymes work by (increasing/decreasing) the amount of energy needed to begin a reaction in the cell.

12. An enzyme acts only on a particular substance that is called a

a. binding site.
b. substrate.
c. complement.
d. histologue.

13. An enzyme's ability to recognize the substance upon which it will act seems to be based on

a. atomic weight.
b. molecular shape or conformation.
c. structural formula.

14. A substance needed to convert an inactive form of an enzyme to an active form is called a(n) _______________ or a _______________.

15. A chemical that interferes with cellular metabolism by denaturing its enzymes is called a(n)

a. antienzyme.
b. poison.
c. caustic.

16. The form of energy utilized by most cellular processes is

a. chemical.
b. electrical.
c. thermal.
d. mechanical.

17. The process by which the energy in a molecule of glucose is released within the cell is called _______________ _______________.

18. The initial phase of respiration that occurs in the cytoplasm and produces 3-carbon pyruvic acid molecules plus energy is called _______________ respiration or _______________.

19. The energy needed to started glycolysis is provided by

a. oxygen.
b. ATP.
c. ADP.
d. cAMP.

20. The second phase of respiration that occurs in the mitochondria and produces carbon dioxide, water, and energy is called _______________ respiration.

21. What element is needed for this second phase to take place? _______________

22. The process in which ADP is converted to ATP is called _______________.

23. Which of the two phases is more important in energy production? _______________

24. Energy released by cellular respiration is stored in

a. mitochondria.
b. glycogen molecules.
c. ATP molecules.
d. muscle cells.

25. A particular sequence of enzyme-controlled reactions leading to the production of particular products is called a _______________ _______________.

26. Muscle fatigue and cramps following strenuous exercise are due to an accumulation of _______________ _______________.

27. The second phase of the metabolic pathway of carbohydrates within the cell is known as the _______________ _______________ _______________ or the _______________ _______________.

28. The function of hydrogen carriers such as nicotinamide adenine dinucleotide (NAD^+) is to

a. act as an oxygen storage molecule.

b. preserve the structure of glucose molecules.

c. assure that the energy in the hydrogen bonds of glucose molecules is retained in a usable form.

d. act as catalysts for oxidation reactions.

29. The main storage reservoirs for glycogen are _______________ cells.

a. liver

b. bone

c. brain

d. muscle

30. What is the fate of carbohydrate that is in excess of the amount that can be stored as glycogen?

a. It is excreted via the kidney.

b. The cells are forced to increase their metabolic rate to burn it.

c. It is converted to fat molecules and deposited in fat tissue.

d. None of the above

31. The substance into which fats must be converted before they can be converted into energy is

a. fatty acids (triglycerides).

b. ketones.

c. glycerol.

32. A person whose fat metabolism is inefficient may develop a "fruity" odor to his or her breath due to an increase in _______________ _______________.

33. If proteins are to be used as an energy source, they must undergo a process in the liver called _______________.

34. What element is removed in this process?

a. carbon

b. hydrogen

c. oxygen

d. nitrogen

35. The rate at which a metabolic pathway functions is regulated by a _______________ _______________.

36. We inherit traits from our parents because

a. DNA contains genes that are the carriers of inheritance.

b. genes tell the cells to construct protein in a unique way for each individual.

c. our species, *Homo sapiens,* reproduces heterosexually.

37. A molecule consisting of a double spiral with sugar and phosphates forming the outer strands and organic bases joining the two strands is _______________.

38. DNA molecules are located in the _______________. Protein synthesis takes place in the _______________.

39. The two types of RNA are _______________ RNA and _______________ RNA.

40. The function of RNA is

a. the formation of lipids, such as cholesterol.

b. to guide the breakdown of polysaccharides.

c. to control the bonding of amino acids.

41. The quantity of a specific protein synthesized by a cell is proportional to the amount of _______________ _______________ that is present.

42. mRNA molecules and ribosomes act as patterns to synthesize

a. protein.

b. carbohydrate.

c. fats.

d. genes.

43. Each codon can specify only one amino acid.

a. True

b. False

44. The amino acids in a molecule of protein are arranged in the correct sequence by

a. ribosomes.

b. transfer RNA.

c. messenger RNA.

d. DNA.

45. The energy for the synthesis of protein molecules is supplied by _______________ molecules.

46. Some antibiotics fight infection by interfering with protein synthesis in bacteria.

a. True

b. False

47. Which of the following factors is capable of causing mutation?

a. radiation

b. poor diet

c. sunlight

d. chemicals

Study Activities

I. Definition of Key Terms and Introduction

A. Define the following key terms used in this chapter.

aerobic respiration

anabolism

anaerobic respiration

catabolism

coenzyme

deamination

dehydration synthesis

DNA

energy

enzyme

gene

genetic code

glycolysis

hydrolysis

metabolism

mutation

oxidation

reduction

RNA

substrate

transcription

translation

B. Describe the cellular mechanisms used by bacterial toxins to affect the human body.

II. Metabolic Processes (pp. 102–104)

A. Answer the following concerning anabolic metabolism.

1. Define *anabolism.*

2. This formula is an example of anabolism. Label the formula and identify the process illustrated.

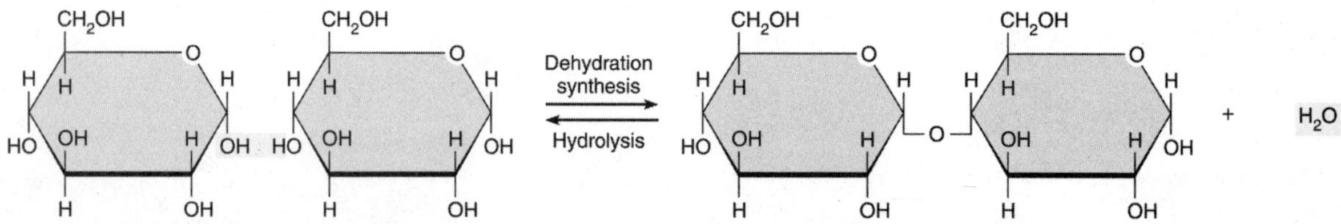

3. Glycerol and fatty acid molecules joined by dehydration synthesis form _______________.
4. Amino acids joined by dehydration synthesis form _______________.

B. Answer these questions concerning catabolism.

1. Define *catabolism.*

2. What process is illustrated here?

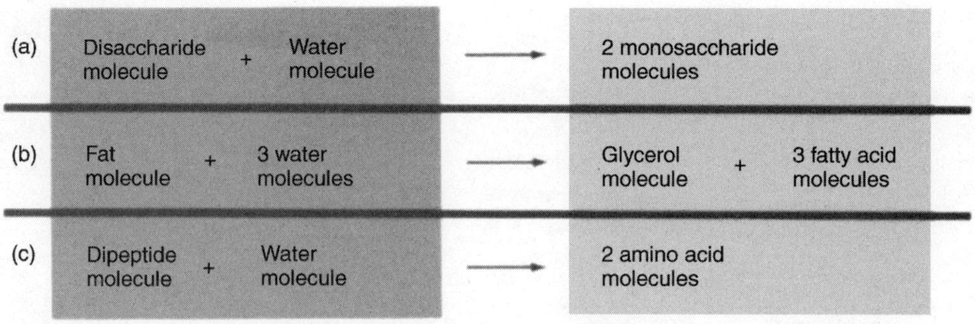

3. Why are anabolism and catabolism balanced in healthy individuals?

III. Control of Metabolic Reactions (pp. 104–105)

A. Describe the composition of enzymes. What is the function of enzymes?

B. Enzymes promote chemical reactions in cells by (increasing/decreasing) the amount of ______________ needed to initiate a reaction.

C. What is the relationship between an enzyme and a substrate? Explain how this works.

D. A sequence of enzyme-controlled reactions is a ______________ ______________.

E. What is a coenzyme? What substances are coenzymes? What is a cofactor? What substances are cofactors?

F. What factors alter enzymes? How will this affect cell function?

G. Explain how the antibiotic penicillin produces its effect.

IV. Energy for Metabolic Reactions (p. 106)

A. What is energy?

B. What are six common forms of energy?

C. What form of energy is used by cell processes?

D. Energy is released by what process?

V. An Overview of Cellular Respiration (pp. 106–108)

A. Fill in the following chart comparing anaerobic and aerobic respiration.

Types of Respiration

Type	Anaerobic	Aerobic
Relative amount of energy used		
End product of oxidation		
Location of reaction within cell		
How released energy is captured		
Number of molecules formed		

B. The primary energy-carrying molecule in the body is ____________.

C. Describe the roles of glycolysis, the citric acid cycle, and oxidative phosphorylation in cellular respiration.

D. Why is the process of forming ATP called oxidative phosphorylation?

VI. A Closer Look at Cellular Respiration (pp. 108–116)

A. What is a metabolic pathway?

B. The metabolic pathways for carbohydrates, lipids, and proteins are such that an excess of any of these substances can be stored as ______________.

C. Carbohydrate pathways

1. Carbohydrates enter ______________ pathways and supply ______________.

2. Describe the citric acid cycle.

3. What happens when glycolysis is blocked?

4. Describe ATP synthesis.

5. How do hydrogen carriers contribute to ATP synthesis?

6. Explain oxygen debt.

7. What is the relationship between lactic acid levels and stress?

8. What is the role of lactic acid in the development of cancer?

9. Where is glucose stored?

D. Lipid pathways

1. Where is lipid metabolism controlled?

2. Describe the lipid pathway.

3. How is an excess of ketone bodies produced? What conditions can produce an excess of ketone bodies?

E. Protein pathways

1. What are the end products of protein metabolism?

2. Eight amino acids are described as essential because the body cannot ______________ them.

F. Describe how the rate of function of a metabolic pathway is regulated.

VII. **Nucleic Acids and Protein Synthesis** (pp. 117–126)

A. Answer these questions concerning genes.

1. What is a gene?

2. The building blocks of nucleic acids are ______________.

3. How are genes necessary to cell metabolism?

B. What is the structure of DNA?

C. How did DNA assist in the identification of Tsar Nicholas II and his family?

D. Answer these questions concerning nucleotides.

1. What are the components of a nucleotide?

2. What are the possible bases of nucleotides in DNA?

E. What is the significance of the sequence of base pairs?

F. Fill in the following chart concerning DNA and RNA.

Comparison of DNA and RNA Molecules

	DNA	RNA
Main location		
6-carbon sugar		
Basic molecular structure		
Organic bases included		
Major functions		

G. Describe the roles of messenger RNA and transfer RNA in protein synthesis. Can messenger and transfer RNA be reused?

H. What are ribosomes and why are they significant?

I. Draw the matching strand of DNA for the strand illustrated here.

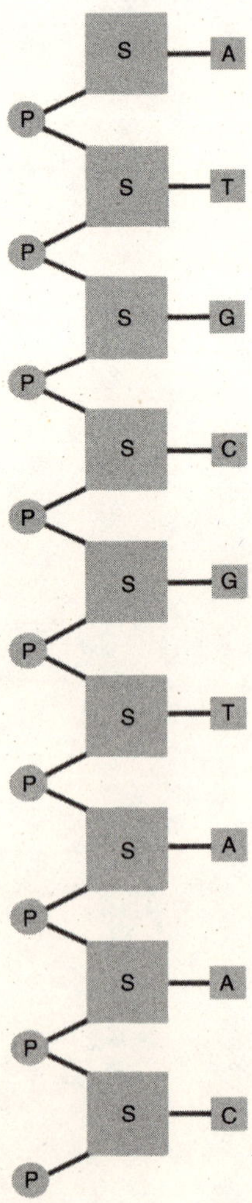

J. What is the energy source for protein synthesis?

K. Describe the role of ribosomes in protein synthesis.

L. How do proteins assume their unique shapes?

M. How are the kinds of proteins produced by the cell controlled?

N. Describe how DNA is replicated.

VIII. Changes in Genetic Information (pp. 126–130)

A. Explain gene amplification and list some clinical applications of this technology.

B. What is a mutation and how does it occur?

C. What can we do to limit the incidence of mutation of our own cells?

IX. The Human Genome Project (pp. 130–131)

What is the Human Genome Project and what are the expected benefits of this project?

X. Clinical Focus Question

Your roommate, a weight lifter, tells you he has a friend from whom he can get drugs to increase his muscle size. What advice would you give him?

When you have completed the study activities to your satisfaction, retake the mastery test and compare your performance with your initial attempt. If there are still areas you do not understand, repeat the appropriate study activities.

CHAPTER 5
TISSUES

OVERVIEW

This chapter deals with the simplest level of organization of cells—tissues. It explains the types of tissues that occur in the human body, the general functions of each of these types of tissues, and organs in which the various types of tissue occur (objectives 1–9).

The characteristics of a tissue remain the same regardless of where it occurs in the body. Knowledge of these characteristics is basic to understanding how a specific tissue contributes to the function of an organ.

CHAPTER OBJECTIVES

After you have studied this chapter, you should be able to:

1. Describe the general characteristics and functions of epithelial tissue.
2. Name the types of epithelium and identify an organ in which each is found.
3. Explain how glands are classified.
4. Describe the general characteristics of connective tissue.
5. Describe the major cell types and fibers of connective tissue.
6. List the types of connective tissues within the body.
7. Describe the major functions of each type of connective tissue.
8. Distinguish among the three types of muscle tissue.
9. Describe the general characteristics and functions of nervous tissue.

FOCUS QUESTION

How is tissue related to the organization of the body?

MASTERY TEST

Now take the mastery test. Do not guess. Some questions may have more than one correct answer. As soon as you complete the test, correct it. Note your successes and failures so that you can read the chapter to meet your learning needs.

1. The Visible Human is available on the Internet.
   a. True
   b. False
2. In complex organisms, cells of similar structure and function are organized into groups called _______________.
3. Which of the following statements about intercellular material is/are true?
   a. The intercellular material is secreted by the tissue cells.
   b. The intercellular substance is living.
   c. The form of intercellular substance is a semisolid gel
4. List the four major tissue types found in the human body.

5. The function of epithelial tissue is to
   a. support body parts.
   b. cover body surfaces.
   c. bind body parts together.
   d. form the framework of organs.

6. Which of the following statements about epithelial tissue is/are true?

a. Epithelial tissue has no blood vessels.

b. Epithelial cells reproduce slowly.

c. Epithelial cells are nourished by substances diffusing from connective tissue.

d. Injuries to epithelial tissue bleed profusely.

7. Cancer cells secrete a substance that dissolves basement membrane tissues.

a. True

b. False

8. Match the following types of epithelial cells with their correct location.

____ 1.	simple squamous epithelium	a.	within columnar or cuboidal epithelium
____ 2.	simple cuboidal epithelium	b.	lining of the ducts of salivary glands
____ 3.	simple columnar epithelium	c.	lining of respiratory passages
____ 4.	pseudostratified columnar epithelium	d.	epidermis of skin
____ 5.	stratified squamous epithelium	e.	air sacs of lungs, walls of capillaries
____ 6.	stratified cuboidal epithelium	f.	lining of digestive tract
____ 7.	stratified columnar epithelium	g.	lining of urinary tract
____ 8.	transitional epithelium	h.	lining of developing ovarian follicle
____ 9.	glandular epithelium	i.	conjunctiva of the eye

9. Glands that secrete their products into ducts are called ______________ glands.

10. The function of connective tissue is

a. support.

b. protection.

c. covering.

d. fat storage.

11. The most common kind of cell in connective tissue is the ______________.

12. A connective tissue cell that can become detached and move about is the ______________.

13. An important characteristic of collagenous fibers is their

a. white color.

b. rigidity.

c. tensile strength.

d. inability to reproduce rapidly.

14. Does statement *a* explain statement *b?*

a. Collagen protein has a very precise configuration.

b. Disruption of collagen protein molecules leads to devastating consequences.

15. Elastic fibers have (more/less) strength than collagenous fibers.

16. Which of the following statements is/are true of connective tissue?

a. Dense connective tissue has many collagen fibers.

b. It contains both elastic and collagenous fibers.

c. Elastin is stronger than collagen.

d. It has a rich blood supply.

17. Sprains heal slowly because dense regular connective tissue has a relatively poor blood supply.

a. True

b. False

18. Which of the following statements is/are true about adipose tissue?

a. It is a specialized form of loose connective tissue.

b. It occurs around the kidneys, behind the eyeballs, and around various joints.

c. It serves as a conserver of body heat.

d. It serves as a storehouse of energy for the body.

19. Tendons connect ______________ to ______________; ligaments connect ______________ to ______________. Both are examples of ______________ connective tissue.

20. The chondromucoprotein is part of the intercellular substance of

    a. muscle.
    b. bone.
    c. cartilage.
    d. nerves.

21. The most rigid connective tissue is ______________.

22. The intercellular material of vascular tissue is ______________.

23. The three types of muscle tissue are ______________, ______________, and ______________.

24. The coordination and regulation of body functions are the functions of ______________ tissue.

25. The cells that bind nerve cells and support nervous tissue are ______________ cells.

Study Activities

I. Definition of Key Terms

Define the following key terms used in this chapter.

adipose tissue

cartilage

chondrocyte

connective tissue

epithelial tissue

fibroblast

fibrous tissue

macrophage

muscle tissue

nervous tissue

neuroglia

neuron

osteocyte

osteon

II. Introduction (p. 136)

A. Describe the Visible Human Project.

B. List the four major tissue types found in the body, and describe their location and function.

III. Epithelial Tissues (pp. 136–44)

A. Where in the body is epithelial tissue present?

B. List the functions of epithelial tissue.

C. Answer the following concerning simple squamous epithelium.

1. Describe the structure of simple squamous epithelium.

2. Simple squamous epithelium is found where ______________ and ______________ take place.

D. Answer the following concerning simple cuboidal epithelium.

1. Describe the structure of simple cuboidal epithelium.

2. Where is this type of tissue found?

3. The function of simple cuboidal epithelium is ______________ and ______________.

E. Answer the following concerning simple columnar epithelium.

1. Describe the structure of simple columnar epithelium.

2. Where does this tissue occur?

3. What is its function?

F. Answer the following concerning pseudostratified columnar epithelium.

1. Microscopic, hairlike projections called ______________ are a characteristic of pseudostratified columnar epithelium.

2. Where is this tissue found?

3. What is its function?

G. Answer the following concerning stratified squamous epithelium.

1. Describe the structure and function of stratified squamous epithelium.

2. Where is this tissue found?

H. Answer these questions concerning stratified cuboidal epithelium.

1. Where is this tissue located?

2. What are its functions?

I. Answer these questions concerning stratified columnar epithelium.

1. Where is this tissue located?

2. What are its functions?

J. Answer these questions concerning transitional epithelium.

1. Where is this tissue located?

2. What are its functions?

K. Describe the location and function of glandular epithelium.

L. A cancer that develops in epithelium is called a _______________.

M. Answer these questions concerning glandular epithelium.

1. Epithelial cells specialized to secrete substances into ducts or body fluids are _______________ _______________.

2. In what type of tissue are such cells usually located?

3. Fill in the following chart.

Types of Exocrine Glands		
Type	**Characteristics**	**Example**
Unicellular glands		
Multicellular glands		
Simple glands		
1. Simple tubular gland		
2. Simple coiled tubular gland		
3. Simple branched tubular gland		
4. Simple branched alveolar gland		
Compound glands		
1. Compound tubular gland		
2. Compound alveolar gland		

4. Answer the following questions about glandular secretions.
   a. Describe the characteristics of the secretions of merocrine, apocrine, and holocrine glands.

   b. Give an example of each of these types of glands.

   c. Differentiate between serous cells and mucous cells.

IV. Connective Tissues (pp. 144–53)

A. What are the functions of connective tissue?

B. Fill in the following chart.

Components of Connective Tissue		
Component	**Characteristics**	**Function**
Fibroblasts		
Macrophages		
Mast cells		
Collagenous fibers (white fibers)		
Elastic fibers (yellow fibers)		
Reticular fibers		

C. How does a disturbance in the ability to synthesize collagen affect the body? Give examples.

D. Describe the structure and function of loose connective tissue.

E. Where is adipose connective tissue found, and what is its function?

F. Describe the structure and function of reticular connective tissue.

G. Answer these questions concerning dense connective tissue.

1. What kind of protein makes up dense connective tissue?

2. What is the difference between a ligament and a tendon?

H. Locate and describe elastic connective tissue.

I. Answer these questions concerning osseous connective tissue.

1. What are the characteristics of osseous connective tissue?

2. Bone injuries heal relatively rapidly. Why is this true?

J. Answer these questions concerning blood.

1. What is the intercellular material of this type of connective tissue?

2. List the types of cells found in plasma.

3. Of the cells found in plasma, which ones function only within the blood vessels?

K. Describe the role of connective tissue in protecting the body against infection.

V. Muscle Tissues (pp. 153–54)

A. What are the characteristics of muscle tissue?

B. Fill in the following chart.

Muscle Tissue

Type	Structure	Control	Location
Skeletal			
Smooth			
Cardiac			

VI. Nervous Tissues (p. 154)

A. What is the basic cell of nervous tissue?

B. What is the function of neuroglial cells in nervous tissue?

C. What is the function of nervous tissue?

VII. Clinical Focus Question

Kenny, age 10, is tall for his age and his large muscles are well developed. He is very anxious to play football with a group of boys who are 14 and 15. His mother asks you if you think this is a good idea. Please include the rationale for your response.

When you have completed the study activities to your satisfaction, retake the mastery test and compare your performance with your initial attempt. If there are still areas you do not understand, repeat the appropriate study activities.

CHAPTER 6
SKIN AND THE INTEGUMENTARY SYSTEM

OVERVIEW

This chapter describes the skin and its appendages. It explains the structure and function of the layers of skin (dermis and epidermis) and the hair, nails, and sweat glands (objectives 1–5). This chapter also explains how the skin helps regulate body temperature and how it responds to environmental factors, such as sunlight and injury (objectives 6 and 7).

Study of the integumentary system is essential to understanding how the body controls interaction between the internal and external environments.

CHAPTER OBJECTIVES

After you have studied this chapter, you should be able to:

1. Describe the four major types of membranes.
2. Describe the structure of the layers of the skin.
3. List the general functions of each layer of the skin.
4. Describe the accessory organs associated with the skin.
5. Explain the functions of each accessory organ.
6. Explain how the skin regulates body temperature.
7. Summarize the factors that determine skin color.

FOCUS QUESTION

You have spent the day on the beach in 90° heat. You return to your air-conditioned home and notice that you have several insect bites that you have scratched open. How does the skin help you to adjust to the changes in temperature and prevent a systemic infection?

MASTERY TEST

Now take the mastery test. Do not guess. Some questions may have more than one correct answer. As soon as you complete the test, correct it. Note your successes and failures so that you can read the chapter to meet your learning needs.

1. Cavities that do not open to the outside of the body are lined with _______________ membranes; the cells of this membrane secrete _______________ fluid.
2. Cavities that open to the outside of the body are lined with _______________ membranes; the cells of these membranes secrete _______________.
3. The three types of tissue that make up synovial membranes are _______________ and _______________; the cells of this membrane secrete _______________ fluid.
4. List the functions of the skin.

5. The outer layer of skin is called the _______________.
6. The inner layer of skin is called the _______________.
7. The masses of connective tissue beneath the inner layers of skin are called the _______________ _______________.
8. The cells of the skin that reproduce are in the
   a. keratin.
   b. stratum corneum.
   c. stratum basale (germinativum).
   d. epidermis.

9. The symptoms of psoriasis are due to

a. increased cell division in the epidermis.
b. increased keratinization of epidermal cells.
c. impaired circulation to the epidermis.
d. separation of the dermal and epidermal layers.

10. Poison ivy is an example of _______________ _______________.

11. Pressure sores are caused primarily by

a. friction wearing off the layers of skin.
b. interruption in the blood supply to areas of skin.
c. inadequate padding over bony prominences.
d. poor nutrition and hydration.

12. The pigment that gives color to the skin is

a. melanin.
b. trichosiderin.
c. biliverdin.
d. bilirubin.

13. Skin cancer is associated with high exposure to _______________.

14. Cutaneous carcinomas arise from _______________ cells.

15. The fingerprints of identical twins are identical.

a. True
b. False

16. The dermis of the skin (contains/does not contain) smooth muscle cells.

17. Blood vessels supplying the skin are located in the

a. dermis.
b. epidermis.
c. subcutaneous layer.

18. The accessory organs of the skin are _______________ _______________, _______________, and _______________ _______________.

19. Hair is composed of _______________ _______________ _______________.

20. The most common skin problem is

a. acne.
b. blisters.
c. contact dermatitis.
d. cancers.

21. The glands usually associated with hair follicles are _______________ glands.

a. apocrine
b. endocrine
c. sebaceous
d. exocrine

22. The active, growing part of the nail is the _______________.

23. The sweat glands associated with the regulation of body temperature are the _______________ glands.

a. endocrine
b. eccrine
c. exocrine
d. apocrine

24. Mammary glands are modified _______________ glands.

25. Which of the following organs and tissues produce the most heat?

a. kidneys
b. bones
c. muscle
d. lungs

26. The primary process by which the body loses heat is

a. radiation.
b. conduction.
c. evaporation.
d. convection.

27. An individual who is genetically incapable of producing melanin is called a(n) _______________.

28. The skin responds to ultraviolet radiation by (increasing/decreasing) the production of melanin.

29. The person with a low concentration of oxygenated blood has a skin condition known as

a. hyperemia.
b. eccymosis.
c. cyanosis.

30. When a cut extends into the dermis, the cells that form the new tissue to hold the edges of the wound together are

a. reticulocytes.
b. fibroblasts.
c. phagocytes.
d. neutrophils.

31. In large, open wounds, the healing process may be accompanied by the formation of _______________.

32. When large areas of the epidermis are destroyed by a burn, the cells are replaced by

a. epithelial cells of hair follicles.
b. the dermis.
c. the stratus germinativum.
d. None of the above

Study Activities

I. Definition of Key Terms and Introduction

A. Define the following key terms used in this chapter.

conduction

convection

cutaneous membrane

dermis

epidermis

evaporation

hair follicle

integumentary

keratinization

melanin

mucous membrane

sebaceous gland

serous membrane

subcutaneous

sweat gland

B. Describe Stevens-Johnson syndrome (toxic epidermal necrolysis).

II. Types of Membranes (p. 161)

Fill in the following chart.

Membranes

Type	Tissues contained	Location	Function
Serous			
Mucous			
Synovial			
Cutaneous			

III. Skin and Its Tissues (pp. 161–68)

A. What kinds of tissue are found in the skin?

B. List the functions of the skin.

C. The layers of the skin are the _______________ and _______________.

D. Answer the following questions concerning diseases of the skin.

1. Describe the causes and effects of the various forms of epidermolysis bullosa.

2. What causes the characteristic lesions of psoriasis?

3. How do allergic contact dermatitis and irritant contact dermatitis differ?

E. Fill in the following chart.

Layers of the epidermis

Layer	Location	Characteristics
Stratum basale		
Stratum spinosum		
Stratum granulosum		
Stratum corneum		

F. Why is water not absorbed through the skin?

G. How are the production of epidermal cells and the loss of stratum corneum related to the development of calluses and corns?

H. Answer the following questions about pressure sores (decubitus ulcers).

1. How does a pressure sore develop?

2. How can the development of pressure sores be prevented?

3. Would you expect a very thin person to be at increased risk for a pressure sore? Why?

I. List the functions of the epidermis.

J. Describe the process by which epidermal cells acquire melanin.

K. Cancer of the skin can arise from the deep layers of the ______________ or from pigmented ______________. Compare the characteristics of the two types of skin cancer.

L. Deep layers of skin are protected from the ultraviolet portion of sunlight by ______________.

M. Describe the structure and function of the dermis.

N. How do the patterns of fingerprints form?

O. How does increased age affect the skin?

P. What types of muscle cells are found in the dermis?

Q. What is the function of nerve tissue in the skin?

R. Describe the structure and function of the subcutaneous layer.

IV. Accessory Organs of the Skin (pp. 168–73)

A. Label the parts of a hair follicle on this illustration: hair shaft, hair follicle, region of cell division, arrector pili muscle, sebaceous gland, keratinized cells, hair papilla, dermal blood vessels, epidermis, dermis, pore, basement membrane, eccrine sweat gland.

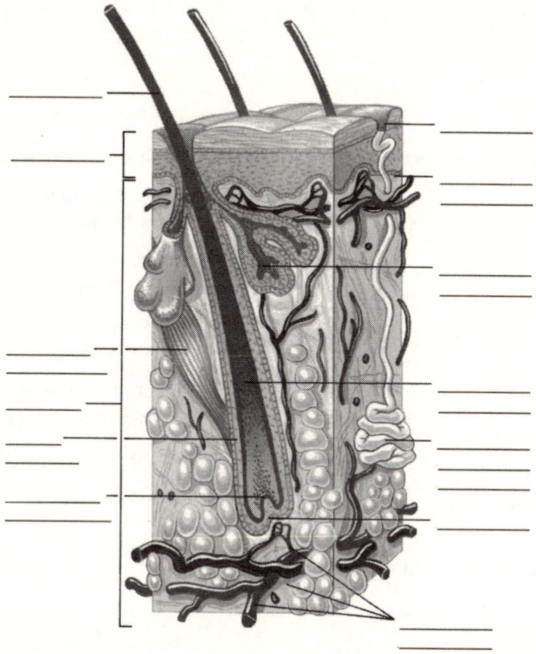

B. Answer the following questions about hair formation.

1. How is hair formed in this follicle?

2. What is folliculitis?

3. Describe several forms of baldness.

4. How is baldness treated? How effective are the various treatments for baldness?

5. Explain how various hair colors are produced.

C. Describe how hair responds to cold temperature or strong emotion.

D. Answer the following questions concerning nails.

1. Where is the growing portion of the nail located?

2. Discuss the clinical significance of the appearance of the nails.

E. Where are sebaceous glands located, and what is the function of the substance they secrete?

F. Compare apocrine and eccrine sweat glands in relation to location, association with other skin structures, and activating stimuli.

V. Regulation of Body Temperature (pp. 173–75)

A. Match the terms in the first column, which are processes of heat loss, with their definitions in the second column.

____ 1. radiation	a. Heat is lost by continuous circulation of air over the body.
____ 2. conduction	b. Sweat changes to vapor and carries heat away from the body.
____ 3. convection	c. Heat moves from the warm body to cooler air via infrared heat waves.
____ 4. evaporation	d. Heat moves away from the body by contact with a cool object.

B. Describe the roles of the nervous, muscular, circulatory, and respiratory systems in heat regulation.

C. Explain the events that can produce an increase in body temperature.

VI. Skin Color (pp. 175–76)

A. Describe the genetic and environmental factors that influence skin color.

B. How do the condition of blood and the blood vessels of the dermis affect skin color?

C. What conditions can produce a yellow color in the skin?

VII. Healing of Wounds and Burns (pp. 177–79)

A. Describe the process of healing in the following wounds.

1. a shallow break in the skin

2. an injury that extends into the dermis or subcutaneous layer

3. a large, open wound

B. Describe the injury and healing process in each of the following conditions.

1. superficial partial-thickness burns

2. deep partial-thickness burns

3. full-thickness burns

VIII. Clinical Focus Question

As summer and the hot weather approach, your family is concerned about your grandmother. Your grandmother is a healthy, 83–year-old woman who lives alone and is proud of her independence. She does not have an air conditioner, despite your family's offer to purchase and install one for her. She says she is too old to get used to newfangled things now. What suggestions could you offer to help your grandmother maintain her health during the hot weather? Explain your rationale.

When you have completed the study activities to your satisfaction, retake the mastery test and compare your performance with your initial attempt. If there are still areas you do not understand, repeat the appropriate study activities.

CHAPTER 7
SKELETAL SYSTEM

OVERVIEW

This chapter deals with the skeletal system—the bones that form the framework for the body. It explains the function and structure of bones, and how they are classified according to structure (objectives 1, 2, and 5). The development of different types of bones and the environmental factors that affect this development (objectives 3 and 4) are also explained. The chapter describes skeletal organization and the location of specific bones within various parts of the skeleton (objectives 6 and 7).

Movement is a characteristic of living things. A study of the skeletal system is necessary to understand how a complex organism, like the human, is organized to accomplish movement.

CHAPTER OBJECTIVES

After you have studied this chapter, you should be able to:

1. Classify bones according to their shapes and name an example from each group.
2. Describe the general structure of a bone and list the functions of its parts.
3. Distinguish between intramembranous and endochondral bones, and explain how such bones grow and develop.
4. Describe the effects of sunlight, nutrition, hormonal secretions, and exercise on bone development.
5. Discuss the major functions of bones.
6. Distinguish between the axial and appendicular skeletons, and name the major parts of each.
7. Locate and identify the bones and the major features of the bones that comprise the skull, vertebral column, thoracic cage, pectoral girdle, upper limb, pelvic girdle, and lower limb.

FOCUS QUESTION

You are playing basketball. Despite your effort to avoid it, the ball strikes you in the head. How has the skeletal system contributed to your ability to move around the court and to protect you from injury?

MASTERY TEST

Now take the mastery test. Do not guess. Some questions have more than one correct answer. As soon as you complete the test, correct it. Note your successes and failures so that you can read the chapter to meet your learning needs.

1. Facial translocation is a complex process that requires close cooperation from a large team of surgeons from various surgical specialties.
   a. True
   b. False
2. Which of the following tissues is found in bones?
   a. cartilage
   b. nerve tissue
   c. fibrous connective tissue
   d. blood
3. A bone with a long longitudinal axis and expanded ends is classified as a _______________ bone.
4. Ribs are examples of _______________ bones.
   a. long
   b. short
   c. flat
   d. sesamoid
5. The shaft of a long bone is the
   a. epiphysis.
   b. diaphysis.

6. The function of a bony process is to provide a

a. passage for blood vessels.

b. place of attachment for tendons and ligaments.

c. smooth surface for articulation with another bone.

d. location for exchange of electrolytes.

7. Which of the following statements about the periosteum is correct?

a. The periosteum contains nerve tissue and is responsible for sensation in bones.

b. The fibers of the periosteum are continuous with ligaments and tendons.

c. The metabolic activity of bone occurs in the periosteum.

d. The periosteum has an important role in bone formation and repair.

8. Bone that consists of tightly packed tissue is called _______________ bone.

9. Bone that consists of numerous bony bars and plates separated by irregular spaces is called _______________ bone.

10. The medullary cavity is filled with

a. spongy bone.

b. fatty connective tissue.

c. marrow.

d. collagen.

11. The intercellular material of bone is _______________ and _______________ _______________.

12. Severe bone pain caused by abnormally shaped red blood cells that obstruct circulation is characteristic of _______________ _______________ disease.

13. Bones that develop from layers of membranous connective tissue are called _______________ _______________.

14. Bones that develop from layers of hyaline cartilage are called _______________ _______________.

15. The band of cartilage between the primary and secondary ossification centers in long bones is called the

a. osteoblastic band.

b. calcium disk.

c. periosteal plate.

d. epiphyseal disk.

16. The presence of the _______________ _______________ on an X ray is an indication that the bone is still growing.

17. Cells undergoing mitosis in the cartilaginous cells of the epiphyseal disk are found in

a. layer one, closest to the epiphyseal end.

b. layer two.

c. layer three.

d. layer four.

18. Which of the following statements about osteoclasts is/are true?

a. Osteoclasts are large cells that originate by the fusion of monocytes.

b. Osteoclasts are cells that give rise to new bone tissue.

c. Osteoclasts become inactive with aging, giving rise to osteoporosis.

d. Osteoclasts get rid of the inorganic component of the oldest cartilaginous cells and allow osteoblasts to invade the region.

19. In a developing bone, compact bone is deposited

a. on the outside of bone just under the periosteum.

b. in the center of the bone within the marrow.

c. on the inner surface of compact bone close to the marrow.

d. in a random fashion within compact bone.

20. Osteoclasts and osteoblasts remodel bone throughout life as osteoclasts resorb bone tissue and osteoblasts replace the bone.

a. True

b. False

21. In bone cancer and in advanced prostatic cancer, the action of osteoclasts becomes dominant.

a. True

b. False

22. Which of the following hormones stimulates cellular activity in the bone tissue?

a. pituitary hormone

b. sex hormones

c. parathyroid hormone

d. thyroid hormone

e. all of the above

23. The effect of exercise on bones is to cause them to _______________, and lack of exercise causes them to _______________.

24. Thyroid hormone primarily affects
   a. intramembranous bone development.
   b. the production of marrow in the medullary cavity of long bones.
   c. the calcification of the epiphyseal disks.
   d. metabolism of bone calcium.

25. A bone fracture that is the result of a disease process is a
   a. pathologic fracture.
   b. greenstick fracture.
   c. closed fracture.
   d. simple fracture.

26. A bone fracture that does not break the skin is called a _______________ fracture.

27. The mass of fibrocartilage that fills the gap between two ends of a broken bone in the early stages of healing is called
   a. a hematoma.
   b. cartilaginous callus.
   c. hyaline cartilage.
   d. granulation tissue.

28. To accomplish movement, bones and muscles function together to act as _______________.

29. In a first-class lever, the pivot is
   a. between the resistance and the force.
   b. before the resistance and the force.
   c. after the resistance and the force.

30. Osteoporosis is characterized by a loss of _______________ volume and _______________ content.

31. Decreased amounts of the hormone _______________ are associated with the development of osteoporosis.

32. The usual number of bones in the human skeleton is _______________.

33. Small bones that develop in tendons where they reduce friction in places where tendons pass over bony prominences are called
   a. sesamoid bones.
   b. irregular bones.
   c. wormian bones.
   d. flat bones.

34. List the four major parts of the axial skeleton.

35. List the four major parts of the appendicular skeleton.

36. The only movable bone of the skull is the
   a. nasal bone.
   b. mandible.
   c. maxilla.
   d. vomer.

37. Air-filled cavities in the cranial bones (sinuses) function to
   a. reduce the weight of the skull.
   b. act as a reservoir for mucus.
   c. control the temperature of structures within the skull.
   d. increase the intensity of the voice by acting as sound chambers.

38. The bone that forms the back of the skull and joins the skull along the lambdoidal suture is the _______________ bone.

39. The bone containing the sella turcica, which protects the pituitary gland, is the _______________ bone.

40. The bones with which all other facial bones articulate are the _______________ bones.

41. A cleft palate is due to incomplete fusion of the _______________ _______________ of the maxilla.

42. The membranous areas (soft spots) of an infant's skull are called _______________.

43. The facial bones that form the orbit of the eye are the _______________ and the _______________ bones.

44. The adult vertebral column has how many parts?

a. 33
b. 23
c. 26
d. 30

45. The intervertebral disks are attached to what part of the vertebrae?

a. lamina
b. body
c. spinous process
d. pedicle

46. A type of vertebral crack or break experienced by such athletes as gymnasts and pole vaulters is a _______________.

47. Which of the vertebrae contain the densest osseous tissue?

a. cervical
b. thoracic
c. lumbar
d. sacral

48. The posterior wall of the pelvic girdle is formed by the _______________.

49. An exaggeration of the thoracic curve is called

a. lordosis.
b. scoliosis.
c. kyphosis.

50. The function of the thoracic cage includes

a. production of blood cells.
b. contribution to breathing.
c. protection of heart and lungs.
d. support of the shoulder girdle.

51. True ribs articulate with _______________ _______________ and the _______________.

52. The middle body of the sternum is the

a. manubrium.
b. tubercle.
c. xiphoid process.
d. body.

53. The union of the manubrium and the body of the sternum is an important anatomic landmark of the chest and is called the _______________ _______________.

54. The pectoral girdle is made up of two _______________ and two _______________.

55. What is commonly referred to as the elbow bone is actually

a. the surgical neck of the humerus.
b. the olecranon process of the ulna.
c. the radial tuberosity.
d. the styloid process.

56. The wrist consists of

a. 8 carpal bones.
b. 5 metacarpal bones.
c. 14 phalanges.
d. distal segments of the radius and the ulna.

57. The bones of the palm of the hand are the _______________ bones.

58. When the hands are placed on the hips, they are placed over

a. the iliac crest.
b. the acetabulum.
c. the ischial tuberosity.
d. the ischial spines.

59. The longest bone in the body is the

a. tibia.
b. fibula.
c. femur.
d. patella.

60. The lower end of the fibula can be felt as an ankle bone. The correct name is the

a. head of the fibula.
b. lateral malleolus.
c. talus.
d. lesser trochanter.

61. The largest of the tarsal bones is the _______________.

Study Activities

I. Definition of Key Terms

Define the following terms used in this chapter.

articular cartilage

compact bone

diaphysis

endochondral bone

epiphyseal disk

epiphysis

fontanel

hematopoiesis

intramembranous bone

lever

marrow

meatus

medullary cavity

osteoblast

osteoclast

osteocyte

osteon

periosteum

spongy bone

trabeculae

II. Introduction (p. 185)

A. What tissues make up bones?

B. Describe facial translocation. Speculate on what kinds of patients may be helped by this kind of surgery.

III. Bone Structure (pp. 185–88)

A. Describe the characteristics of each of the following types of bone: long bones, short bones, flat bones, irregular bones, and sesamoid bones.

B. Answer the following questions about the parts of a long bone.

1. The expanded articular part of a long bone is called the ______________.
2. The articulating surface is coated with a layer of ______________ ______________.
3. The shaft of a long bone is known as its ______________.
4. Describe the periosteum and its functions.

5. How is the shape of a bone related to its function?

6. Describe compact and spongy bone, and their functions.

C. Label the following parts in this drawing of a long bone: proximal epiphysis, diaphysis, distal epiphysis, articular cartilage, spongy bone, space occupied by red marrow, compact bone, medullary cavity, yellow marrow, periosteum, epiphyseal disks.

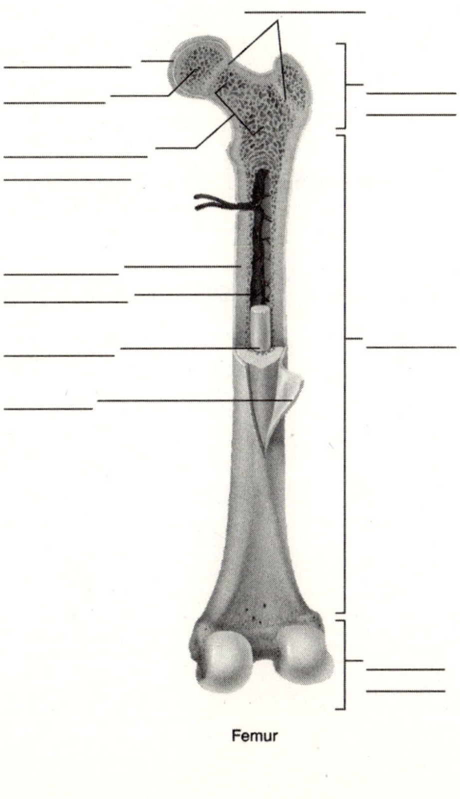

Femur

D. Answer the following concerning the microscopic structure of bone.

1. Bone cells (osteocytes) are located in _______________, which are arranged in concentric circles around _______________ or _______________ canals.

2. What are the intracellular materials of bone? What are the functions of these materials?

3. Describe the blood supply of bone.

4. What are the structural differences in compact bone and spongy bone?

5. Why do patients who suffer from sickle cell disease have bone pain?

IV. Bone Development and Growth (pp. 188–93)

A. What bones are intramembranous bones? How do these develop?

B. What bones are endochondral bones? How do these develop? Be sure to include descriptions of the primary ossification center, the secondary ossification center, and the epiphyseal disk.

C. Answer these questions concerning ossification.

1. When can ossification centers first be detected on an X ray?

2. When is ossification complete?

3. How can an X ray show that long bone growth is complete?

4. Why do men who have prostatic cancer have new bone formation on the bony trabeculae?

D. Fill in the following chart.

Factors influencing bone development

Factors	**Effect on bone development**
Vitamin D	
Vitamin A *Excess and deficiency*	
Vitamin C	
Growth hormone *Excess and deficiency*	
Thyroid hormone	
Male and female sex hormones	
Physical exercise	

E. What is rickets?

F. Pituitary dwarfism is treated with _______________ _______________ _______________.

G. Describe the events in healing a fracture from rupture of the periosteum to formation of a bony callus.

H. What factors influence the rate at which a fracture heals?

V. Bone Function (pp. 193–99)

A. What bones function primarily to provide support?

B. What bones function primarily to protect viscera?

C. Answer the following concerning levers.

1. Identify each of these three types of levers.

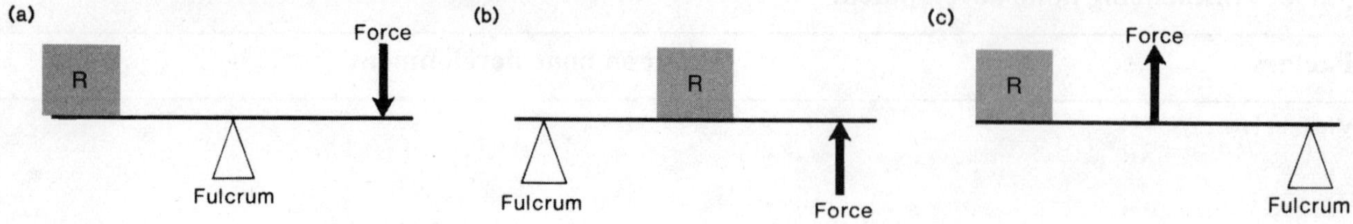

2. When does the arm function as a first-class lever? A third-class lever?

D. Answer these questions concerning blood cell formation.

1. Where are blood cells formed in the embryo? In the infant? In the adult?

2. What is the difference between red and yellow marrow?

E. Answer these questions concerning inorganic compounds in bone.

1. What are the major inorganic salts stored in bone? What other salts and heavy metals can also be stored in bone?

2. How is calcium released from bone so that it is available for physiological processes?

3. What physiological processes depend on calcium ions in the blood?

4. Excessive loss of bone volume and mineral content associated with aging is ______________.

5. What are the effects of this process? How is this condition diagnosed?

6. What measures are recommended to prevent osteoporosis?

VI. Skeletal Organization (pp. 199–201)

A. The adult skeleton usually contains _____ bones. Why does this number vary?

B. What are the two major divisions of the skeleton?

C. List the bones found in each of these major divisions.

VII. **Skull** (pp. 214 220)

A. Answer these questions concerning the number of bones in the skull.

1. How many bones are found in the human skull?

2. How many of these bones are found in the cranium?

3. How many in the facial skeleton?

B. Answer the following questions concerning the cranial bones.

1. Using your own head or that of a partner, locate the following cranial bones and identify the suture lines that form their boundaries: occipital bone, temporal bones, frontal bone, and parietal bones.

2. What are the remaining two bones of the cranium? Where are they located?

C. Answer the following concerning the facial bones.

1. Using yourself or a partner, locate the following facial bones: maxilla, palatine, zygomatic, lacrimal bones, nasal bones, vomer bone, inferior nasal conchae, and mandible.

2. Which of the facial bones is known as the keystone of the face? Why?

3. Which of the facial bones is the only movable bone of the skull?

4. How does a cleft palate develop?

5. Describe the differences between the infant skull and the adult skull.

D. Fill in the following chart.

Passageways through the bones of the skull

Passageway	Location	Major structures transmitted
Carotid canal		
Foramen lacerum		
Foramen magnum		
Foramen ovale		
Foramen rotundum		
Foramen spinosum		
Greater palatine foramen		
Hypoglossal canal		
Incisive foramen		
Inferior orbital tissue		
Infraorbital foramen		
Internal acoustic meatus		
Jugular foramen		
Mandibular foramen		
Mental foramen		
Optic canal		
Stylomastoid foramen		
Superior orbital fissure		
Supraorbital foramen		

VIII. Vertebral Column (pp. 214–20)

A. Answer the following questions about the vertebral column.

1. What is the function of the vertebral column?

2. Draw the normal curves of the vertebral column in the margin of this page.

3. What is the difference between the vertebral column of an infant and that of an adult? How does this occur?

B. Label the parts of the accompanying diagram: lamina, spinous process, transverse process, pedicle, vertebral foramen, superior articulating process, body.

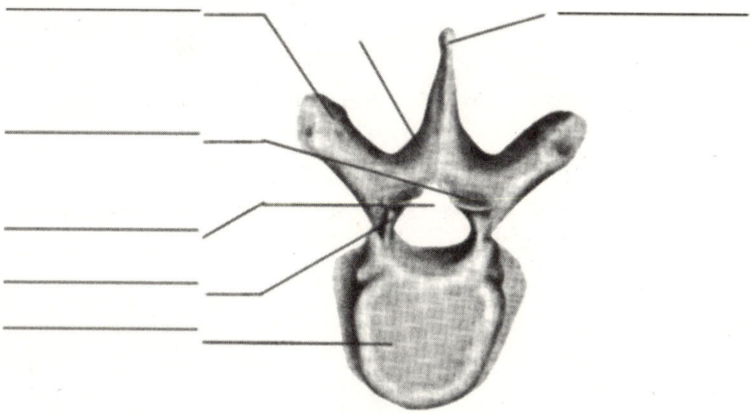

C. In what ways is the structure of the thoracic vertebrae unique?

D. In what ways is the structure of the lumbar vertebrae unique?

E. What is the importance of the sacrum in obstetrics?

F. Answer the following questions concerning problems of the vertebral column.

1. Describe the abnormal spinal curves. Be sure to include problems associated with these abnormal curves.

2. What happens when an intervertebral disk ruptures or herniates?

3. What changes occur in the vertebral column as a result of aging?

IX. Thoracic Cage (pp. 220–22)

A. Name the bones of the thoracic cage.

B. Describe the differences among true, false, and floating ribs. Include their articulations.

C. Describe the sternum including manubrium, body, sternal angle, and xiphoid process. Locate these structures on yourself or a partner.

D. What is a sternal puncture and why is it done?

X. Pectoral Girdle (pp. 222–23)

Use yourself or a partner to locate and list the bones of the pectoral girdle. What is the function of the pectoral girdle?

XI. Upper Limb (pp. 223–27)

A. Use yourself or a partner to locate and list the bones of the upper limb.

B. Label these parts in the following drawing: phalanges, metacarpals, carpals, pisiform, triquetrum, hamate, lunate, capitate, scaphoid, trapezoid, trapezium, distal phalanx, middle phalanx, proximal phalanx.

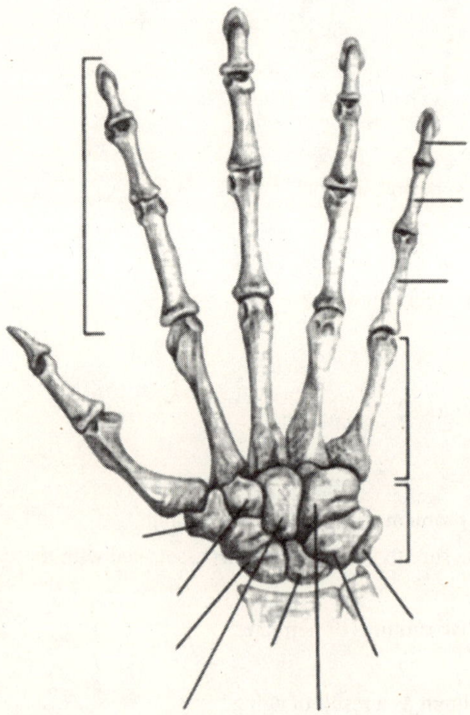

XII. **Pelvic Girdle** (pp. 227–31)

A. List the bones of the pelvic girdle.

B. Identify the bone in which each of the structures in the following chart is located, and explain the function of each structure.

Structures of the bones of the pelvic girdle and their functions

Structure	Bone	Function
Acetabulum		
Anterior superior iliac spine		
Symphysis pubis		
Obturator foramen		
Ischial tuberosity		
Ischial spines		

C. Describe the differences between the male pelvis and the female pelvis.

XIII. **Lower Limb** (pp. 231–35)

A. List the bones of the lower limb.

B. Identify the bone in which each of the structures in the following chart is located, and explain the function of each structure.

Structures of the bones of the lower limb and their functions

Structure	Bone	Function
Fovea capitis		
Medial malleolus		
Lateral malleolus		
Greater and lesser trochanters		
Tibial tuberosity		

C. Describe patellar dislocation and explain how it can be prevented.

D. Label these structures on the following illustration: tarsals, metatarsals, phalanges, calcaneus, talus, navicular, cuboid, lateral cuneiform, intermediate cuneiform, medial cuneiform, proximal phalanx, middle phalanx, distal phalanx.

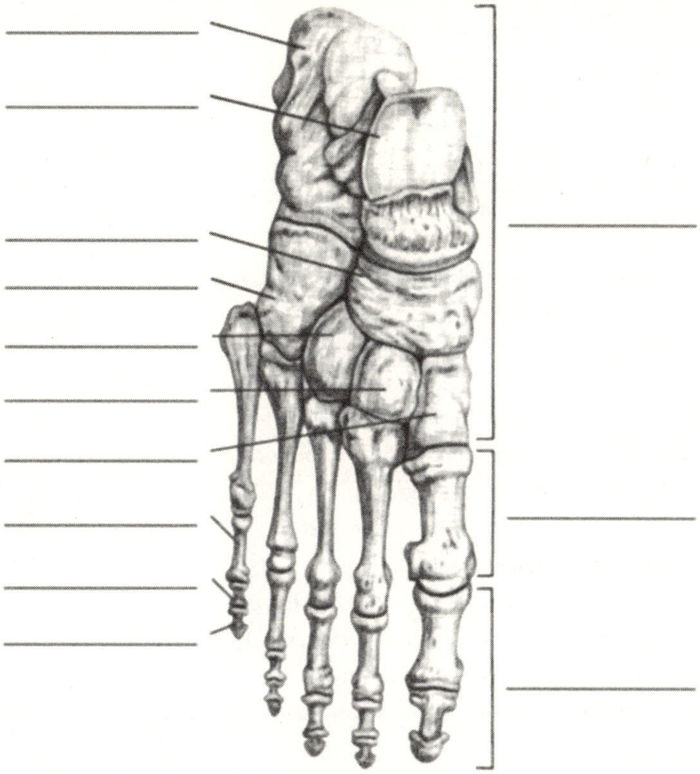

E. How is the foot able to support the weight of the body?

XIV. Clinical Focus Question

Your neighbors' 2-week-old infant has been diagnosed as having a mild congenital hip displacement, and the doctor has told the parents that the use of a thick diaper should correct the problem. Both parents are very upset and state that they do not understand what is wrong with the baby or the reason for the heavy diaper. How would you explain the diagnosis and treatment to them?

When you have completed the study activities to your satisfaction, retake the mastery test and compare your performance with your initial attempt. If there are still areas you do not understand, repeat the appropriate study activities.

CHAPTER 8
JOINTS OF THE SKELETAL SYSTEM

OVERVIEW

This chapter describes the ways in which the bones of the skeletal system are joined together to form joints (objectives 1–3, 7 & 8). You will study how the joints are classified and how the various kinds of joints are held together (objectives 4 & 5). The role of the skeletal muscles in producing movement and the various kinds of movement possible are described (objective 6).

An understanding of how joints work is basic to understanding how the body moves.

CHAPTER OBJECTIVES

After you have studied this chapter, you should be able to:

1. Explain how joints can be classified according to the type of tissue that binds the bones together.
2. Describe how bones of fibrous joints are held together.
3. Describe how bones of cartilaginous joints are held together.
4. Describe the general structure of a synovial joint.
5. List six types of synovial joints and name an example of each type.
6. Explain how skeletal muscles produce movements at joints and identify several types of joint movements.
7. Describe the shoulder joint and explain how its articulating parts are held together.
8. Describe the elbow, hip, and knee joints and explain how their articulating parts are held together.

FOCUS QUESTION

You finish transcribing your class notes, rise from your chair, and stretch. How do the joints enable you to perform these movements?

MASTERY TEST

Now take the mastery test. Do not guess. Some questions may have more than one correct answer. As soon as you complete the test, correct it. Note your successes and failures so that you can read the chapter to meet your learning needs.

1. In humans, the joints most often affected by gout (accumulations of uric acid crystals) are
   a. ball and socket joints like the shoulder.
   b. saddle joints like the thumb.
   c. hinge joints like the elbow.
   d. hinge joints like the big toe.
2. The function of joints is to
   a. bind parts of the skeletal system together.
   b. allow movement in response to skeletal muscle contraction.
   c. permit bone growth.
   d. all of the above
3. Name three classifications of joints according to movement and according to the type of tissue that binds them together.

4. Which of the following are characteristics of fibrous joints?
   a. The bones of the joint have a space between them.
   b. The bones of the joint are held firmly together by fibrous connective tissue.
   c. This type of joint is found in the skull.
   d. The structure of these joints is fixed early in life.
5. Syndesmosis, suture, and gomphosis are types of _______________ joints.

6. The epiphyseal disk is an example of a _______________ or a _______________ _______________.

7. Movement in a vertebral column and the symphysis pubis (is/is not) due to compressing a pad of cartilage.

8. Diarthroses are (more/less) complex in structure than synarthroses and amphiarthroses.

9. The function of articular cartilage is to

   a. provide flexibility in the joint.
   b. provide insulation.
   c. minimize friction.
   d. secrete synovial fluid.

10. Shock absorption in a synovial joint is the function of the _______________ _______________.

11. A joint condition that is secondary to stress such as obesity or injury is

   a. suchondral plate fracture.
   b. degenerative arthritis.
   c. rheumatoid arthritis.
   d. bursitis.

12. If aspirated synovial fluid is red-tinged and contains pus, the most likely cause is

   a. a fracture.
   b. osteoarthritis.
   c. gout.
   d. a bacterial infection.
   e. a collagen disease, rheumatoid arthritis.

13. The joint structures that limit movement in a joint in order to prevent injury are the

   articulating surfaces of the bones.
   ligaments.
   c. tendons.
   d. synovial membranes.

14. The inner layer of the joint capsule is the _______________ _______________.

15. Which of the following are functions of synovial fluid?

   a. lubrication of the joint surfaces
   b. prevention of infection within the joint
   c. nutrition of the cartilage within the joint
   d. absorption of shock within the joint

16. Disks of fibrocartilage within a joint that help distribute body weight within the joint are called _______________.

17. A fluid-filled sac within a joint is a _______________.

18. Articular cartilage receives its supply of oxygen and nutrients from _______________ _______________.

19. The type of joint that permits the widest range of motion is a _______________ joint.

   a. pivot
   b. hinge
   c. gliding
   d. ball-and-socket

20. Match the joint in the first column with the type of joint it represents.

   ____ 1. shoulder
   ____ 2. elbow
   ____ 3. ankle
   ____ 4. thumb

   a. saddle
   b. gliding
   c. ball-and-socket
   d. pivot
   e. hinge

21. The two bones that form the shoulder joint are the _______________ and the _______________.

22. The shoulder (is/is not) an extremely stable joint.

23. The kind of injury to which the shoulder joint is prone is _______________.

24. The _______________ and the _______________ make up the hinge joint of the elbow.

25. What movements are made possible by the rotation of the head of the radius?

26. An instrument used to visualize the interior of a joint is the _______________.

27. The head of the femur fits into the _______________ of the _______________ bone.

28. Following joint replacement surgery, the joint must be kept at rest for an extended period of time so that full range of motion may be regained.

   a. True
   b. False

29. List the six possible movements of the hip joint.

30. The largest and most complex of the synovial joints is the ______________ joint.

31. Rotation at the knee joint is possible when the knee is

a. flexed.

b. extended.

c. abducted.

d. adducted.

33. A joint injury that involves stretching and tearing of ligaments and tendons is a ______________.

Study Activities

I. Definition of Key Terms

Define the following terms used in this chapter.

articulation

bursa

gomphosis

ligament

meniscus

suture

symphysis

synchondrosis

syndesmosis

synovial

II. Introduction (p. 257)

A. The place where two or more bones meet is called a ______________ or an ______________.

B. List the functions of joints.

III. Classification of Joints (pp. 257–60)

A. Fibrous joints (synarthroses)

1. List the characteristics of fibrous joints.

2. Describe and give an example of each of the following fibrous joints.

   a. syndesmosis

   b. suture

c. gomphosis

3. Areas in the infant skull that permit the shape of the skull to change during childbirth are called _______________.

4. Why can sutures be used to estimate the age of the skull?

B. Cartilaginous joints (amphiarthroses)

1. List and describe the two types of cartilaginous joints.

2. Label the following parts of the drawings of two cartilaginous joints: intervertebral disk, body of vertebra, gelatinous core, band of fibrocartilage, pubic bone, fibrocartilaginous disk of the symphysis pubis.

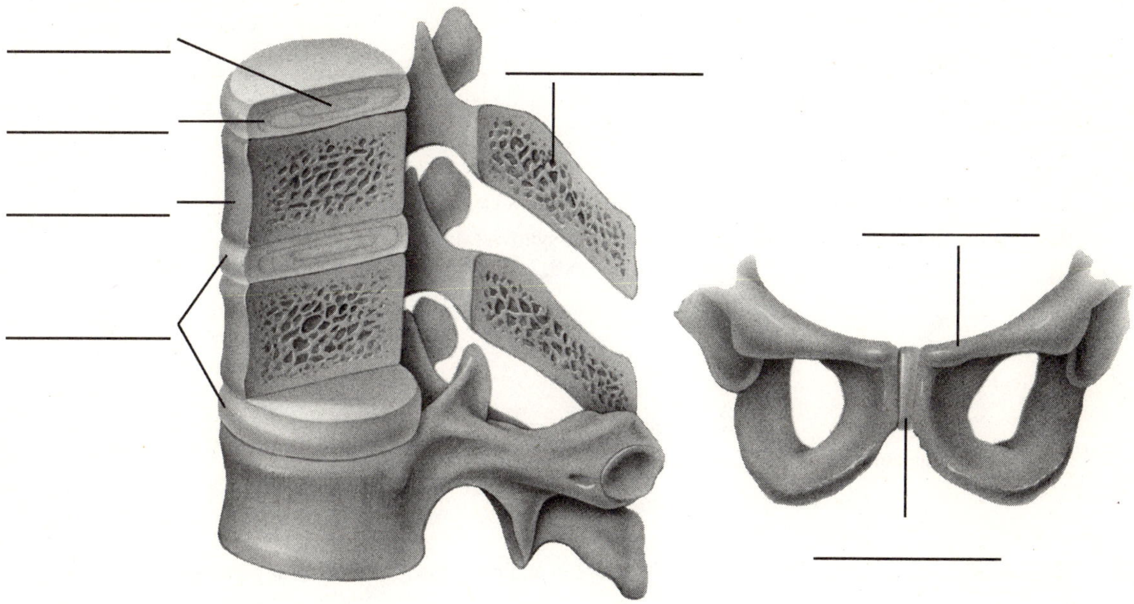

3. Describe the function of an intervertebral disk.

4. A symphysis important in childbirth is the _______________ _______________.

IV. General Structure of a Synovial Joint (pp. 260–62)

A. Label the following parts of a synovial joint: subchondral plate, joint cavity filled with synovial fluid, spongy bone, joint capsule, articular cartilage, synovial membrane.

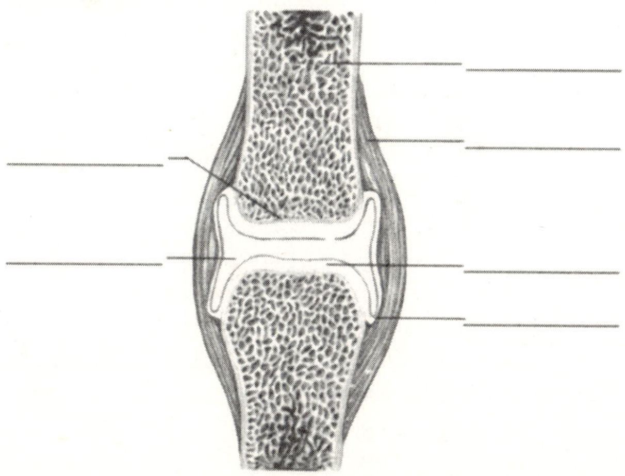

B. List the conditions that can be diagnosed by examining synovial fluid. Include the clinical findings for each condition.

C. Answer the following questions about the structure and function of synovial joints.

1. The parts of bones that come together in a joint are covered by a layer of _______________ _______________.
2. Articular cartilage lies on the subchondral plate, which usually contains _______________ bone.
3. What is the function of the subchondral plate?
4. Describe osteoarthritis. What factors lead to this condition?
5. The bones of a diarthrosis are held together by a _______________ _______________.
6. The outer layer of the structure in the previous question is composed of _______________ connective tissue and is attached to the _______________ of the bone.
7. _______________ bind the articular ends of bone together.
8. The membrane that covers all surfaces within the joint capsule is the _______________ _______________.
9. List the functions of the synovial membrane and synovial fluid.
10. Describe the menisci and their function.
11. Describe the bursae and their function.
12. How does immobility affect the supply of oxygen and nutrients to the articular cartilage?

V. Types of Synovial Joints (p. 262)

Complete the following chart related to synovial joints.

Type	Description	Possible movement	Example
Ball-and-socket			
Condyloid			
Gliding			
Hinge			
Pivot			
Saddle			

VI. Types of Joint Movements (pp. 262–67)

Describe the following joint movements. You may also wish to perform these movements as you describe them.

flexion

extension

hyperextension

dorsiflexion

plantar flexion

abduction

adduction

rotation

circumduction

supination

pronation

eversion

inversion

protraction

retraction

elevation

depression

VII. Examples of Synovial Joints (pp. 268–75)

A. Shoulder joint

1. Label the following parts of the shoulder joint: clavicle, acromion of scapula, transverse humeral ligament, humerus, scapula, subscapular bursa, articular capsule, coracoid process of scapula.

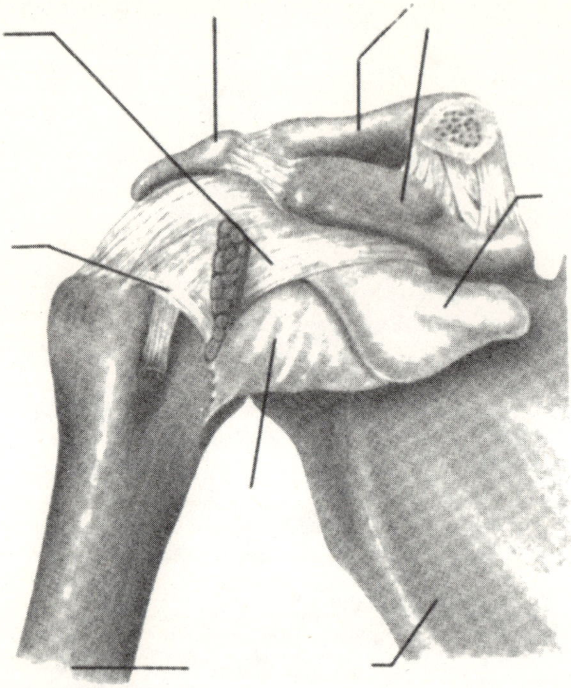

2. What is the rotator cuff? How is this structure injured?

3. Explain the relationship between the wide range of movement at the shoulder joint and the relative ease with which the shoulder can be dislocated.

4. List the ligaments that help prevent shoulder dislocation.

B. Elbow joint

1. Describe the structure of the elbow joint.

2. What types of movement are possible at this joint?

3. The procedure used to diagnose and treat injuries to the elbow, shoulder, and knee via a thin fiber-optic instrument is called _______________.

4. What condition is diagnosed using polymerase chain reaction?

C. Hip joint

1. The hip joint is a _______________ _______________ joint.

2. List the structures of the hip joint and describe their functions.

3. The hip joint is (more/less) movable than the shoulder joint. Give the rationale for your answer.

4. Describe the procedure for arthroplasty.

5. List the major ligaments of the hip and identify the function of each.

6. Describe the procedure for hip joint replacement.

D. Knee joint

1. Label the following parts of the knee joint: femur, synovial membrane, suprapatellar bursa, patella, prepatellar bursa, joint cavity, articular cartilages, meniscus, infrapatellar bursa, joint capsule, tibia.

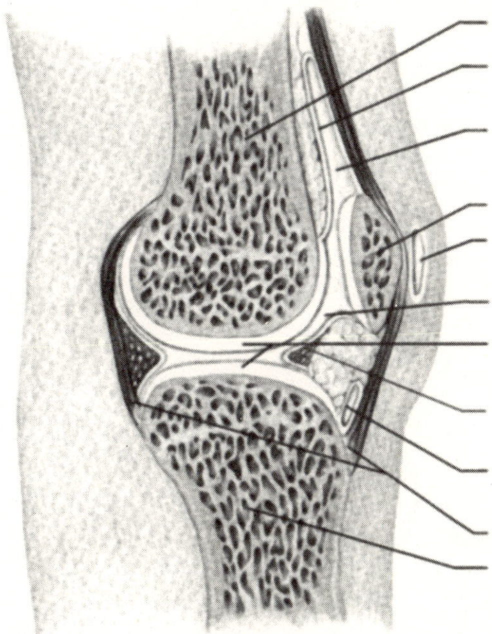

2. List and describe the five major ligaments of the knee joint.

3. What is the function of the cruciate ligaments?

4. What happens when one of these ligaments is torn?

5. What are the results of damaging a meniscus in the knee?

6. List the movements possible in the knee joint.

7. If the knee is distended above and along the sides of the patella, one would suspect _______________ _______________.

VIII. Joint Disorders (pp. 276–77)

A. Tearing or overstretching the connective tissues, ligaments, and tendons associated with a joint due to a forceful wrenching or twisting is called a _______________.

B. Inflammation of a bursa due to excessive use of a joint is called _______________.

C. Compare the clinical features of rheumatoid arthritis and osteoarthritis.

D. List the types of arthritis that are caused by infections.

IX. Clinical Focus Question

After fracturing your humerus just distal to the surgical neck, your arm was immobilized in a sling that bound your upper arm to your trunk for six weeks. The sling has just been removed and your physician has prescribed physical therapy for you. Why was your arm immobilized in this fashion? What kinds of exercises can you anticipate the physical therapist will prescribe for you?

When you have completed the study activities to your satisfaction, retake the mastery test and compare your performance with your initial attempt. If there are still areas you do not understand, repeat the appropriate study activities.

CHAPTER 9
MUSCULAR SYSTEM

OVERVIEW

This chapter presents the muscular system. In conjunction with the skeletal system, the muscular system serves to move the body. The chapter introduces the three types of muscle; the major events in contraction of skeletal, smooth, and cardiac muscles; the energy supply to muscle fiber for contraction; the occurrence of oxygen debt; and the process of muscle fatigue (objectives 1, 3, 4, and 10). The chapter describes the structure and function of a skeletal muscle, distinguishes between a twitch and a sustained contraction, explains how various kinds of muscle contraction produce body movements and maintain posture, shows how the location and interaction of muscles produce body movements, identifies the location and action of major skeletal muscles, and differentiates the structure and function of a multiunit smooth muscle and a visceral smooth muscle (objectives 2, 5-9, 11, and 12).

The skeletal system can be thought of as the passive partner in producing movement; the muscular system can be thought of as the active partner. This chapter explains how muscles interact with bones to maintain posture and produce movement. In addition, it tells the characteristics and functions of skeletal, smooth, and cardiac muscles. This knowledge is a foundation for the study of other organ systems, such as the digestive system, the respiratory system, and the cardiovascular system.

CHAPTER OBJECTIVES

After you have studied this chapter, you should be able to:

1. Describe how connective tissue is part of the structure of a skeletal muscle.
2. Name the major parts of a skeletal muscle fiber and describe the function of each part.
3. Explain the major events that occur during muscle fiber contraction.
4. Explain how energy is supplied to the muscle fiber contraction mechanism, how oxygen debt develops, and how a muscle may become fatigued.
5. Distinguish between fast and slow muscle fibers.
6. Distinguish between a twitch and a sustained contraction.
7. Describe how exercise affects skeletal muscles.
8. Explain how various types of muscular contractions produce body movements and help maintain posture.
9. Distinguish between the structures and functions of a multiunit smooth muscle and a visceral smooth muscle.
10. Compare the contraction mechanisms of skeletal, smooth, and cardiac muscle fibers.
11. Explain how the locations of skeletal muscles help produce movements and how muscles interact.
12. Identify and locate the major skeletal muscles of each body region and describe the action of each muscle.

FOCUS QUESTION

How do muscle cells utilize energy and interact with bones to accomplish such diverse movements as playing a piano and playing a basketball game?

MASTERY TEST

Now take the mastery test. Do not guess. Some questions may have more than one correct answer. As soon as you complete the test, correct it. Note your successes and failures so that you can read the chapter to meet your learning needs.

1. A skeletal muscle is covered with fibrous connective tissue called
   a. fascia.
   b. aponeuroses.
   c. perimysium.
   d. sarcolemma.

2. Layers of connective tissue that attach muscle to the periosteum are called
   a. ligaments.
   b. tendons.
   c. aponeuroses.
   d. elastin.
3. Bundles of muscle fibers are called _______________.
4. Each bundle is covered by layers of connective tissue called _______________.
5. The connective tissue that divides the individual muscles into compartments is the
   a. deep fascia.
   b. superficial fascia.
   c. epimysium.
   d. perimysium.
6. A surgical procedure to relieve pressure within a muscle compartment is a _______________.
7. The characteristic striated appearance of skeletal muscle is due to the arrangement of alternating protein filaments composed of _______________ and _______________.
8. The area between two Z lines in a myofibril is called a(n)
   a. transverse tubule.
   b. sarcomere.
   c. actin filament.
   d. myosin filament.
9. Striated muscle tissue is also known as _______________ muscle.
   a. voluntary
   b. involuntary
   c. skeletal
   d. visceral
10. When muscle fibers are overstretched, the injury sustained is a _______________ _______________.
11. The union between a nerve fiber and a muscle fiber is the
    a. motor neuron.
    b. motor end plate.
    c. neuromuscular junction.
    d. neurotransmitter.
12. At this time, the cause of post-polio syndrome is thought to be
    a. reactivation of past infection with the polio virus.
    b. a new infection of susceptible nerve cells.
    c. accelerated aging of previously damaged nerve cells.
    d. unknown.
13. The response of muscle tissue to stimulation is _______________.
14. The precision of movement produced by a muscle is due to
    a. the size of the muscle fiber, small fibers being more precise.
    b. the small muscle fiber–to–neuron ratio within a motor unit.
    c. many muscle fibers being present for each neuron in a motor unit.
    d. the number of branches in the neuron, many branches being associated with diffuse stimulation.
15. The finer and more precise the movement produced by a particular muscle, the (fewer/greater) the number of muscle fibers in the motor unit.
16. The most abundant protein in muscle is _______________.
17. A defect in the myosin cross-bridge of cardiac muscle that slows the rate at which actin slides past myosin is known as _______________ _______________ cardiomyopathy.
18. When the filaments of actin and myosin merge within the myofibril, the result is
    a. shortening of the muscle fiber.
    b. membrane polarization.
    c. release of acetylcholine.
19. Duchenne muscular dystrophy is due to the absence of the muscle protein _______________.
20. The role of acetylcholine in membrane depolarization is to _______________ the muscle fiber membrane.
21. What ion is necessary in relatively high concentrations to allow the formation of cross-bridges between actin and myosin?
22. When a muscle fiber is at rest, the protein complex _______________ prevents the formation of cross-bridges between actin and myosin.

23. The above protein complex is rendered inactive by the presence of

a. acetylcholine.
b. calcium.
c. anticholinesterase.
d. sodium.

24. The substance that halts the stimulation of a muscle fiber by a neuron is _______________.

25. A toxin that can prevent the release of the neurotransmitter from motor nerve fibers is produced by the bacterium _______________ _______________.

26. The energy used in muscle contraction is supplied by the decomposition of _______________ _______________.

27. The primary source of energy to reconstruct ATP from ADP and phosphate is a substance called _______________ _______________.

28. A person feels out of breath after vigorous exercise because of oxygen debt. Which of the following statements helps explain this phenomenon?

a. Anaerobic respiration increases during strenuous activity.
b. Lactic acid is metabolized more efficiently when the body is at rest.
c. Conversion of lactic acid to glycogen occurs in the liver and requires energy.
d. Priority in energy use is given to ATP synthesis.

29. After prolonged muscle use, muscle fatigue occurs due to an accumulation of _______________ _______________.

30. The partial contraction of skeletal muscles after death is called _______________ _______________.

31. The rate of speed at which various muscles contract is related to

a. the number of muscle fibers in the motor unit.
b. the amount of myoglobin present in muscle tissue.
c. the number of mitochondria present in the cytoplasm of the muscle cells.
d. differences in cell membrane permeability in various muscle cells.

32. When muscles are active, large amounts of _______________ are released.

33. The minimal strength stimulus needed to elicit contraction of a single muscle fiber is called a _______________ _______________.

34. The strength of a muscle contraction in response to different levels of stimulation is determined by

a. the level of stimulation delivered to individual muscle fibers.
b. the number of fibers that respond in each motor unit.
c. the number of motor units stimulated.
d. the characteristics of each muscle group.

35. A record of a muscle's response to stimulation is called a _______________.

36. The period of time following a muscle response to a stimulus when it will not respond to a second stimulus is called the

a. latent period.
b. contraction.
c. refractory period.

37. Muscle tone refers to

a. a state of sustained, partial contraction of muscles that is necessary to maintain posture.
b. a feeling of well-being following exercise.
c. the ability of a muscle to maintain contraction against an outside force.
d. the condition athletes attain after intensive training.

38. When exercise activates primarily the slow, red fibers, the result is

a. increased muscle strength.
b. increased muscle size.
c. increased resistance to fatigue.
d. increased anaerobic tolerance.

39. Smooth muscle contracts (more slowly/more rapidly) than skeletal muscle following stimulation.

40. Two types of smooth muscle are _______________ muscle and _______________ muscle.

41. Peristalsis is due to which of the following characteristics of smooth muscle?

a. the capacity of smooth muscle fibers to excite each other
b. automaticity
c. rhythmicity
d. sympathetic innervation

42. The self-exciting property of cardiac muscle is probably due to

a. the presence of intercalated disks between muscle cells.

b. a well-developed sarcoplasmic reticulum.

c. a cell membrane more permeable to sodium ions.

d. the presence of increased amounts of nonionized calcium.

43. In the following statements, does statement *a* explain statement *b?* _______________

a. Cardiac muscle remains refractory until a contraction is completed.

b. Sustained tetanic contraction is not possible in heart muscle.

44. Drugs that stop spasms in cardiac muscle by interfering with the movement of calcium ions across the cardiac muscle cell membranes are called _______________ _______________ _______________.

45. The attachment of a muscle to a relatively fixed part is called the _______________; the attachment to a relatively movable part is called the _______________.

46. Smooth body movements depend on _______________ giving way to prime movers.

47. The muscle that compresses the cheeks inward when it contracts is the

a. orbicularis oris.

b. epicranius.

c. platysma.

d. buccinator.

48. Excessive use of jaw muscles to clench the jaw may lead to _______________ _______________ syndrome.

49. The muscle that moves the head so that the face turns to the opposite side when one side contracts is the

a. sternocleidomastoid.

b. splenius capitis.

c. semispinalis capitis.

d. longissimus capitis.

50. The muscle that abducts the upper arm and can both flex and extend the humerus is the

a. biceps brachii.

b. deltoid.

c. infraspinatus.

d. triceps brachii.

51. The muscle that extends the arm at the elbow is the

a. biceps brachii.

b. brachialis.

c. supinator.

d. triceps brachii.

52. The band of tough connective tissue that extends from the xiphoid process to the symphysis pubis and serves as an attachment for muscles of the abdominal wall is the _______________ _______________.

53. The heaviest muscle in the body, which serves to straighten the leg at the hip during walking, is the

a. psoas major.

b. gluteus maximus.

c. adductor longus.

d. gracilis.

54. The tendon that connects the gastrocnemius muscle with the calcaneus is the _______________ tendon, or the _______________ tendon.

Study Activities

I. Definition of Key Terms

Define the following key terms used in this chapter.

actin

antagonist

aponeurosis

fascia

insertion

motor neuron

muscle impulse

myofibril

myogram

myosin

neurotransmitter

origin

oxygen debt

prime mover

recruitment

sarcomere

synergist

threshold stimulus

II. **Introduction** (p. 282)

A. What are the three types of muscle?

B. What are the pathologic mechanisms and treatment of Moebius syndrome?

III. **Structure of a Skeletal Muscle** (pp. 282–87)

A. List the kinds of tissue present in skeletal muscle.

B. A skeletal muscle is held in position by layers of fibrous connective tissue called _______________. This tissue extends beyond the end of a skeletal muscle to form a cordlike _______________. When this tissue extends beyond the muscle to form a sheetlike structure, it is called a(n) _______________.

C. List those joints most commonly affected by tendonitis and tenosynovitis.

D. Label these parts in the following drawings of skeletal muscle: bone, tendon, fascia, skeletal muscle, epimysium, perimysium, fascicle, endomysium, muscle fibers, sarcolemma, sarcoplasmic reticulum, filaments, endomysium, nucleus, myofibrils, blood vessels, axon of motor neuron.

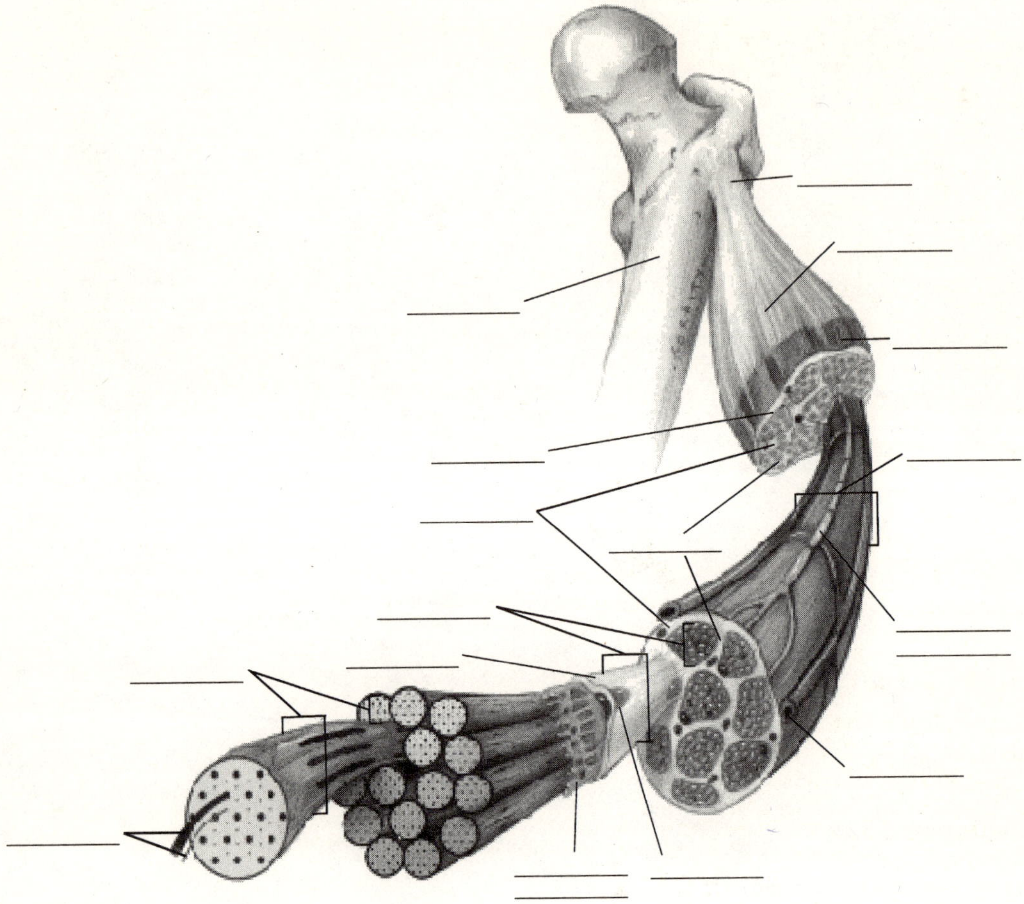

E. Answer the following questions about fascia.

1. Where are deep, subcutaneous, and subserous fascia located?

2. What is a fasciotomy and why might it be performed?

F. Answer the following concerning skeletal muscle fibers.

1. The contractile unit of a muscle is a _______________ _______________ _______________.

2. Describe the structure and function of a sarcomere.

3. The network of membranous channels in the cytoplasm of muscle fibers is the _______________ _______________.

4. What is the function of these channels?

G. Describe the injuries found in a muscle strain.

H. What is post-polio syndrome?

I. Label these structures in the accompanying illustration of a single fiber motor unit: mitochondria, synaptic cleft, synaptic vesicles, folded sarcolemma, muscle fiber nucleus, myofibril of muscle fiber, motor end plate, nerve fiber branches, motor neuron fiber.

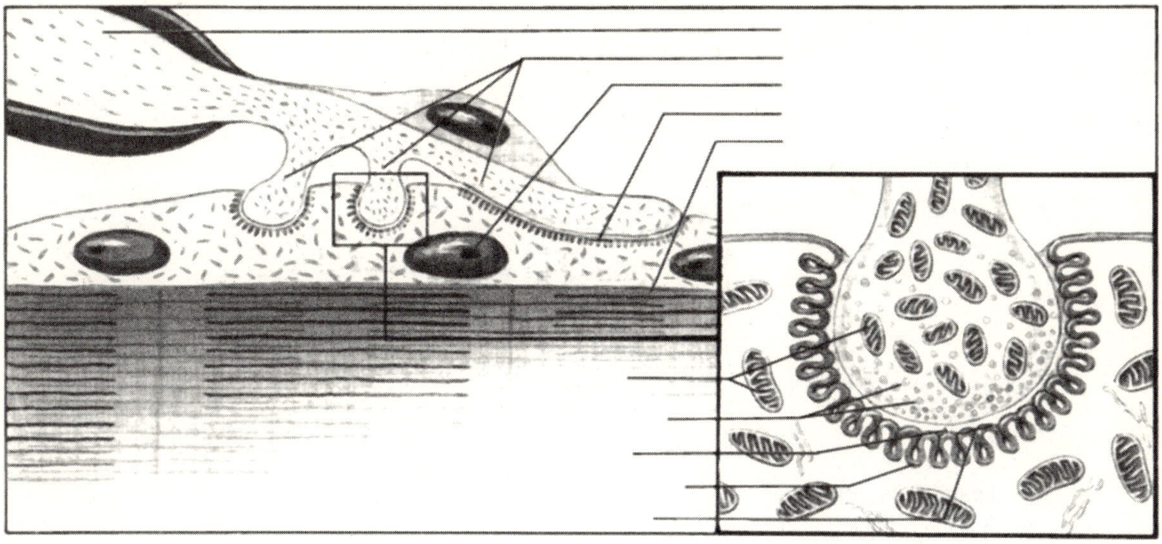

J. Describe how the strength of a muscle contraction and the precision of the movement produced are determined.

IV. Skeletal Muscle Contraction (pp. 287–94)

A. Describe the roles of actin, myosin, tropomyosin, and troponin in muscle contraction.

B. What is the role of dystrophin in skeletal muscle function?

C. Answer the following concerning stimulus for contraction.

1. Describe the interaction of acetylcholine and calcium ions in stimulating muscle contraction.

2. The action of acetylcholine is halted by the enzyme _______________.

3. What is the action of botulinus toxin?

4. What is the result of ingesting this toxin?

D. Answer these questions concerning energy sources for contraction.

1. How does ATP supply energy for muscle contraction?

2. How does creatine phosphate supply energy for muscle contraction?

E. Answer these questions concerning oxygen supply and cellular respiration.

1. What substance in muscle seems able to store oxygen temporarily?

2. How is oxygen usually brought to muscle cells?

3. Why is oxygen necessary for muscle contraction?

4. How does muscle continue to contract in the absence of oxygen?

5. What is meant by oxygen debt? How is it paid off?

F. Answer the following concerning muscle fatigue.

1. What is meant by muscle fatigue? What causes it?

2. How do athletes decrease their muscle fatigue?

3. About 25% of the energy released by cellular respiration is available for metabolic processes. About 75% is lost as _______________.

4. Describe the process of rigor mortis.

G. Describe the differences between fast and slow muscles.

V. **Muscular Responses** (pp. 294–97)

A. Answer the following concerning muscular responses.

1. Define *threshold stimulus.*

2. Define *all-or-none response.*

3. How is all-or-none response related to the strength of a muscle contraction?

B. The accompanying illustration shows a myogram of a type of muscle contraction known as a twitch. Label the following: time of stimulation, latent period, period of contraction, period of relaxation, refractory period. Explain the significance of each of these events.

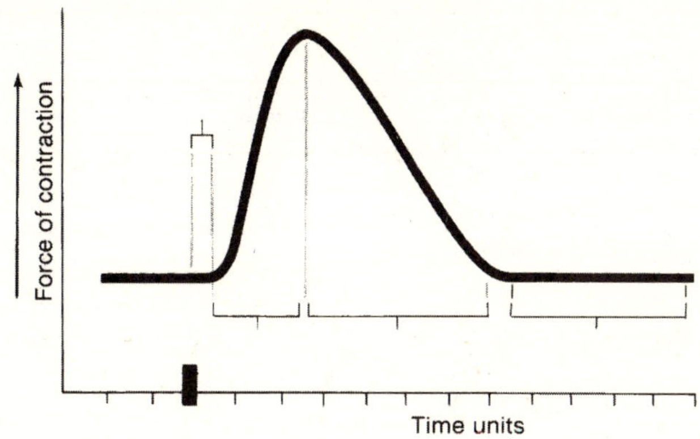

C. Describe these kinds of muscle contractions.

sustained contraction

tetanic contraction

muscle tone

isometric contraction

isotonic contraction

D. Compare the effects of exercise such as swimming and running with weightlifting. Be sure to note the kinds of muscle fibers involved.

E. Complete the following chart.

Changes in muscle size

Change	Conditions leading to change	Observable change in muscle
Hypertrophy		
Atrophy		

VI. Smooth and Cardiac Muscles (pp. 297–99)

Complete the following chart.

Types of muscle tissue

	Skeletal	Smooth	Cardiac
Major location			
Major function			
Cellular characteristics			
striations			
nucleus			
special features			
Mode of control			
Contraction characteristics			

VII. Skeletal Muscle Actions (p. 300)

A. A skeletal muscle has at least two places of attachment to bone. For instance, the gluteus maximus, which extends the leg at the hip, is attached to the posterior surface of the ilium, the sacrum, and the coccyx at one end and to the posterior surface of the femur and the iliotibial tract at the other. One place of attachment is the origin and the other is the insertion. Explain the difference between the two.

B. Match the terms in the first column with the statements in the second column that best describe the role of muscle groups in producing smooth muscle movement.

____ 1. prime mover

____ 2. Synergist

____ 3. antagonist

a. muscle that returns a part to its original position

b. muscle that makes the action of the prime mover more effective

c. muscle that has the major responsibility for producing a movement

VIII. Major Skeletal Muscles (pp. 301–31)

A. List and locate the muscles of facial expression.

B. Answer the following about the muscles of mastication.

1. List and locate the muscles of mastication.

2. Describe TMJ syndrome.

C. List and locate the muscles that move the head.

D. List and locate the muscles that move the upper arm.

E. List and locate the muscles that move the forearm.

F. List and locate the muscles that move the wrist, hand, and fingers.

G. List and locate the muscles of the abdominal wall.

H. List and locate the muscles of the pelvic outlet.

I. List and locate the muscles that move the thigh.

J. List and locate the muscles that move the lower leg.

K. List and locate the muscles that move the ankle, foot, and toes.

IX. Clinical Focus Question

Your sister, age 28, is very excited about the opening of a gym in her neighborhood. Many of her friends have joined and she is planning to get in shape. She had her second child three months ago and has been experiencing difficulty losing weight. Your sister is known in the family as a couch potato. What advice will you give her about her plans? Explain your rationale.

When you have completed the study activities to your satisfaction, retake the mastery test and compare your performance with your initial attempt. If there are still areas you do not understand, repeat the appropriate study activities.

CHAPTER 10

NERVOUS SYSTEM I: BASIC STRUCTURE AND FUNCTION

OVERVIEW

The human body uses two systems to coordinate and integrate the functions of the other body systems so that the internal environment remains stable. These systems are the nervous system and the endocrine system. Chapters 10 and 11 focus on the nervous system (objective 1). Chapter 10 concentrates on the structure and function of the nervous system at the cellular level (objectives 2, 3, 5, and 6–11). Regeneration of injured nerve fibers is discussed (objective 4), and reflex behavior is explained (objectives 12 and 13).

Knowledge of the cellular components of the nervous system is necessary for understanding how the nervous system coordinates body functions.

CHAPTER OBJECTIVES

After you have studied this chapter, you should be able to:

1. Explain the general functions of the nervous system.
2. Describe the general structure of a neuron.
3. Explain how neurons are classified.
4. Name four types of neuroglial cells and describe the functions of each.
5. Explain how an injured nerve fiber may regenerate.
6. Explain how a membrane becomes polarized.
7. Describe the events that lead to the conduction of a nerve impulse.
8. Explain how a nerve impulse is transmitted from one neuron to another.
9. Distinguish between excitatory and inhibitory postsynaptic potentials.
10. Explain two ways impulses are processed in neuronal pools.

FOCUS QUESTION

How is the nervous system organized at the cellular level to coordinate and integrate the functions of the other body systems?

MASTERY TEST

Now take the mastery test. Do not guess. Some questions may have more than one correct answer. As soon as you complete the test, correct it. Note your successes and failures so that you can read the chapter to meet your learning needs.

1. What protein forms plaques in the brains of people who have Alzheimer's disease?
   a. amyloid.
   b. actin.
   c. acetylcholine.
   d. myelin.
2. Improved ability to diagnose Alzheimer's disease is important because
   a. treatment can begin early in the disease.
   b. families can begin to plan for the care of their Alzheimer's patient.
   c. Alzheimer's can be differentiated from other forms of dementia more accurately.
   d. adult children of Alzheimer's patients may use the information in making decisions about childbearing.
3. The two basic types of cells found in neural tissue are _______________ and _______________ cells.
4. The small spaces between neurons are called _________________.

5. The nervous system is composed of two groups of organs called the _______________ nervous system and the _______________ nervous system.

6. Monitoring such phenomena as light, sound, and temperature is a _______________ function of the nervous system.

7. The basic unit of structure and function of the nervous system is the _______________.

8. Another name for the cell body of a neuron is the _______________ or _______________.

9. To what organelle in non-neural cells do Nissl bodies in neurons correspond?
   a. mitochondria
   b. endoplasmic reticulum
   c. nucleii
   d. lysosomes

10. Which of the following structures is *not* common to all nerve cells?
   a. cell body
   b. axon
   c. dendrite
   d. Schwann cells

11. Does statement *a* explain statement *b?* _______________
   a. The nucleus of the nerve cell seems incapable of mitosis.
   b. The nerve cell cannot reproduce.

12. Migraine headaches are thought to be the result of
   a. contraction following by dilation of blood vessels in the brain.
   b. increased levels of norepinephrine in the brain.
   c. increased, uncontrolled electrical activity in the brain.
   d. inflammation of cranial nerves.

13. The structure that carries impulses away from the cell body of the neuron is the
   a. dendrite.
   b. neurofibril.
   c. axon.
   d. neurilemma.

14. The neurilemma is composed of
   a. Nissl bodies.
   b. myelin.
   c. the cytoplasm and nucleus of Schwann cells.
   d. neuron cell bodies.

15. The symptoms of multiple sclerosis are due to
   a. degeneration of myelin.
   b. degeneration of the neural cell membrane.
   c. fatty deposits around dendrites and axons.
   d. a viral infection.

16. The type of neuron that lies totally within the central nervous system is the
   a. sensory neuron.
   b. motor neuron.
   c. interneuron.
   d. unipolar neuron.

17. The supporting framework of the nervous system is composed of
   a. neurons.
   b. dendrites.
   c. neuroglial cells.
   d. myelin.

18. The neuroglial cells that can phagocytize bacterial cells and increase when there is inflammation of the brain or spinal cord are
   a. astrocytes.
   b. oligodendrocytes.
   c. microglia.
   d. ependyma.

19. All but one of the following are functions of the neuroglial cells.
   a. fill spaces
   b. support growth of neurons
   c. hold organs together
   d. phagocytize bacteria

20. The relative immaturity of an infant's nervous system compared to an adult's nervous system is due to
   a. fewer neurologic cells in an infant.
   b. fewer neurons in the infant nervous system compared to the adult nervous system.
   c. incomplete myelinization of the infant nervous system.
   d. poorly developed sensory receptors.

21. Which of the following injuries to nervous tissue can be repaired?

a. damage to a cell body
b. damage to nerve fibers that have myelin sheaths
c. damage to nerve fibers that have a neurilemma
d. Nerve damage cannot be repaired.

22. A tangled mass of regenerating nerve fibers that can be quite painful when sensory fibers are involved is called a ______________.

23. Which of the following accurately reflects the order of ease with which ions pass through nerve cell membranes?

a. potassium, sodium, calcium
b. calcium, potassium, sodium
c. sodium, calcium, potassium
d. potassium, calcium, sodium

24. The difference in electrical charge between the inside and the outside of the membrane in the resting nerve cell is called the ______________ ______________.

25. The effect of a new stimulus of the same type being received before the effect of the previous stimulus in a series has subsided is called ______________.

26. The propagation of action potentials along a fiber is called

a. a threshold potential.
b. repolarization.
c. a nerve impulse.
d. a sensation.

27. The period of total depolarization of the neuron membrane when the neuron cannot respond to a second stimulus is called the ______________ ______________ period.

28. The refractory period acts to limit the

a. intensity of nerve impulses.
b. rate of conduction of nerve impulses.
c. permeability of nerve cell membranes.
d. responsiveness of receptor cells.

29. In which type of fiber is conduction faster?

a. myelinated
b. unmyelinated

30. A decrease in calcium ions below normal limits will

a. facilitate the movement of sodium across the cell membrane.
b. inhibit the movement of sodium across the cell membrane.
c. facilitate the movement of potassium across the cell membrane.
d. inhibit the movement of potassium across the cell membrane.

31. Which of the following statements best explains the ability of procaine to produce local anesthesia?

a. Procaine binds calcium, thereby decreasing the amount of ionized calcium.
b. Procaine decreases the membrane permeability to sodium.
c. Procaine enhances the movement of potassium across the cell membrane.
d. Procaine blocks the release of acetylcholine.

32. The neurotransmitter that stimulates the contraction of skeletal muscles is

a. dopamine.
b. acetylcholine.
c. gamma-aminobutyric acid.
d. encephalins.

33. The amount of neurotransmitter released at a synapse is controlled by

a. calcium.
b. sodium.
c. potassium.
d. magnesium.

34. Continuous stimulation of a neuron on the distal side of this junction is prevented by

a. exhaustion of the nerve fiber.
b. the chemical instability of neurotransmitters.
c. enzymes within the neural junction.
d. rapid depletion of ionized calcium.

35. Neuropeptides that are synthesized by the brain and spinal cord in response to pain are ______________.

36. If a neuron receives a subthreshold stimulus that renders it more likely to respond to incoming excitatory stimulation than before, it is said to be ______________.

37. The process that allows coordination of incoming impulses that represent information from a variety of receptors is called ______________.

38. The most common neuron structure is one

a. axon and many dendrites.
b. process that serves as both axon and dendrite.
c. dendrite and many axons.
d. dendrite and one axon.

39. Unipolar neurons are found in the

a. brain.
b. spinal cord.
c. special sense organs.
d. ganglia.

40. Neurons may be classified functionally as _______________, _______________, and _______________ neurons.

Study Activities

I. Definition of Key Terms

Define the following key terms used in this chapter.

action potential

axon

central nervous system

convergence

dendrite

divergence

effector

facilitation

myelin

neurilemma

neuroglia

neuron

neurotransmitter

Nissl body

peripheral nervous system

receptor

reflex

summation

synapse

threshold

II. Introduction (p. 344)

How can genetic research hold the potential for effective treatment of Alzheimer's disease?

III. General Functions of the Nervous System (pp. 344–49)

A. List the organs of the central and peripheral nervous systems.

B. What are the three general functions of the nervous system?

C. Answer the following concerning the nervous system.

1. Describe the sensory function of the nervous system.

2. Describe the motor function of the nervous system.

D. Answer the following questions about migraines.

1. A migraine is a response to changes in the ______________ ______________ of the face, head, and neck.
2. Define
   a. classic migraine

   b. common migraine

3. What is the treatment for migraine headaches?

E. Answer the following concerning neuron structure.

1. Label these structures in the following drawing of motor neurons: axon, nucleolus, Nissl bodies, cell body, nucleus, neurofibrils, collateral, nucleus of Schwann cell, nodes of Ranvier, myelin, dendrites, axonal hillock.

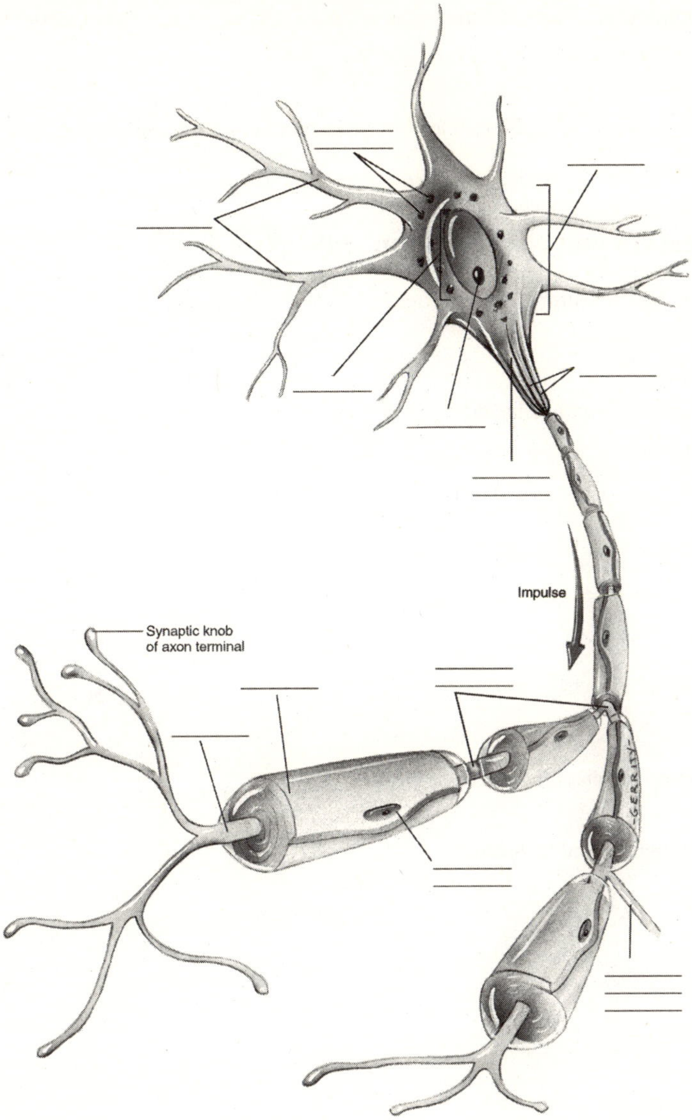

2. How are neurons cultured in the laboratory?

3. Match the parts of a neuron in the first column with the correct description in the second column.

____ 1. neurofibrils

____ 2. Nissl bodies

____ 3. dendrites

____ 4. axon

____ 5. Schwann cells

a. slender fiber that carries impulses away from the cell body; this fiber may give off collaterals

b. membranous sacs in the cytoplasm associated with the manufacture of protein molecules

c. cells of the myelin sheath

d. network of fine threads that extends into nerve fibers

e. short, branched fibers that carry impulses away from the cell body as well as transporting certain organelles

4. Describe how Schwann cells make up the myelin sheath and the neurilemma on the outsides of nerve fibers.

5. What is the composition of white matter in the brain and spinal cord? Of gray matter?

F. Describe the pathologic events of multiple sclerosis, including possible causes of the disease.

IV. Classification of Neurons and Neuroglia (pp. 349–54)

A. What are two ways in which neurons are classified?

B. Describe each kind of neuron and its location.

bipolar

unipolar

multipolar

C. The neurons in section *B* are classified according to _______________.

D. Describe each kind of neuron and its location.

sensory neuron

interneuron

motor neuron

accelerator

inhibitor

E. Fill in the following chart.

Neuroglial cells			
Cell	**Location**	**Structure**	**Function**
Astrocytes			
Microglia			
Oligodendrocytes			
Ependyma			

F. How is myelinization related to the growth and development of the human body?

G. Describe the regeneration of nerve fibers. Include a description of a neuroma.

H. What is Tay-Sachs disease?

V. Cell Membrane Potential (pp. 354–60)

A. Describe membrane potential, resting potential, and action potential. Which of these events is a nerve impulse?

B. What happens when a threshold potential is reached?

C. How would hyperpolarization affect the threshold potential?

D. Answer the following concerning impulse conduction.

1. How do the nodes of Ranvier affect nerve impulse conduction? What kind of conduction is this called?

2. Define *refractory period, absolute refractory period, relative refractory period,* and *all-or-none response* in neurons.

3. What is the result of increasing the permeability of the cell membrane to sodium? Of decreasing the permeability?

4. How does calcium affect nerve impulse conduction?

5. Procaine and cocaine decrease membrane permeability to ______________ ions and thus (increase/decrease) nerve impulse conduction.

VI. The Synapse (pp. 360–65)

A. Answer the following concerning synaptic transmission.

1. Label these structures in the accompanying drawing of a synapse: axon, synaptic knob, mitochondrion, synaptic vesicles, synaptic clefts, dendrite, axon membrane, neurotransmitter substance, polarized membrane, vesicle releasing neurotransmitter, depolarized membrane.

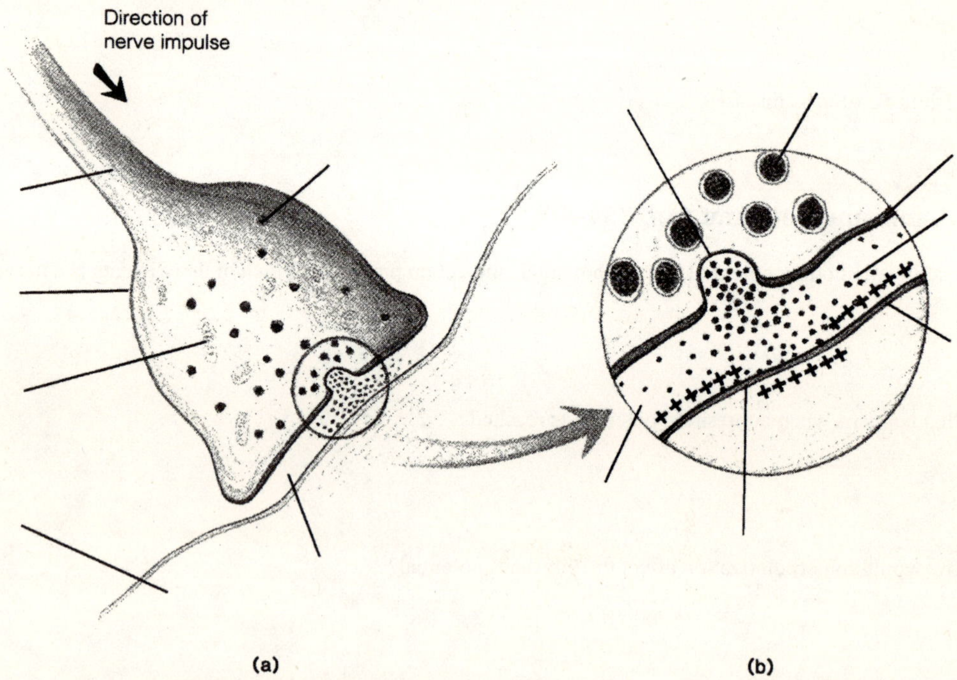

2. How does a neurotransmitter initiate depolarization? (Include the role of both the presynaptic and the postsynaptic neuron membranes.)

3. How is stimulation of a nerve fiber stopped?

4. List substances that function as neurotransmitters.

5. What is the function of neuropeptides?

B. Answer these questions concerning excitatory and inhibitory actions.

1. Describe synaptic potentials, excitatory postsynaptic potentials, and inhibitory postsynaptic potentials. How do they interact in normal nerve function?

2. What substance seems to have inhibitory action?

C. What is the effect of myasthenia gravis on synaptic transmission? What kinds of drugs are helpful in treating this disease?

D. Describe the role of endorphins in the nervous system.

VII. **Impulse Processing** (pp. 365–69)

A. What is a neuronal pool?

B. Explain the relationship between neuronal pools and facilitation, convergence and divergence.

C. Answer the following questions about addiction and the role of receptors.

1. Briefly describe the history of addiction.

2. Define addiction.

3. Describe the role of neurotransmitters and their receptors in the development of addiction.

4. List the symptoms of nicotine addiction.

VIII. Clinical Focus Question

Jack, age 24, amputated his finger while cutting the lawn two days ago. The amputated part was brought to the hospital and reattached using microsurgery techniques. Jack is very angry as you meet him this morning. He explains that the surgeon was quite sure that the surgery would be successful but he still has no sensation in the finger. What would you tell Jack?

When you have completed the study activities to your satisfaction, retake the mastery test. If there are still some areas you do not understand, repeat the appropriate study activities.

CHAPTER 11
NERVOUS SYSTEM II: DIVISIONS OF THE NERVOUS SYSTEM

OVERVIEW

This chapter continues the study of the nervous system. It includes the structure and function of the central nervous system and its coverings (objectives 1–8), the structure and function of the peripheral nervous system (objectives 9–13), and the structure and function of the autonomic nervous system (objectives 14–17).

A knowledge of the component parts of the nervous system is essential to understanding the effects of injury on the body.

CHAPTER OBJECTIVES

After you have studied this chapter, you should be able to:

1. Describe the coverings of the brain and spinal cord.
2. Describe the structure of the spinal cord and its major functions.
3. Describe a reflex arc.
4. Define reflex behavior.
5. Name the major parts of the brain and describe the functions of each.
6. Distinguish among motor, sensory, and association areas of the cerebral cortex.
7. Explain hemisphere dominance.
8. Explain the stages in memory storage.
9. Describe the formation and function of cerebrospinal fluid.
10. Explain the functions of the limbic system and the reticular formation.
11. List the major parts of the peripheral nervous system.
12. Describe the structure of a peripheral nerve.
13. Name the cranial nerves and list their major functions.
14. Explain how spinal nerves are named and their functions.
15. Describe the general characteristics of the autonomic nervous system.
16. Distinguish between the sympathetic and the parasympathetic divisions of the autonomic nervous system.
17. Describe a sympathetic and a parasympathetic nerve pathway.
18. Explain how the autonomic neurotransmitters differently affect visceral effectors.

FOCUS QUESTION

It is noon, and you are just finishing an anatomy assignment. You hear your stomach growling and you realize you are hungry. You make a ham sandwich and pour a glass of milk. After eating, you decide you have been studying for three hours and you should go for a walk. How does the nervous system receive internal and external cues, process incoming information, and decide what action to take?

MASTERY TEST

Now take the mastery test. Do not guess. Some questions have more than one correct answer. As soon as you complete the test, correct it. Note your successes and failures so that you can read the chapter to meet your learning needs.

1. The organs of the central nervous system are the _______________ and the _______________ _______________.

2. The outer membrane covering the brain is composed of fibrous connective tissues and is called the

a. dura mater.
b. arachnoid mater.
c. pia mater.
d. periosteum.

3. A collection of blood under the dura mater secondary to injury to the head is a _______________ _______________.

4. Cerebrospinal fluid is found between the

a. arachnoid mater and the dura mater.
b. vertebrae and the meninges.
c. pia mater and the arachnoid mater.

5. Meningitis is most likely to involve inflammation of the

a. dura mater.
b. arachnoid mater.
c. pia mater.

6. The spinal cord ends

a. at the sacrum.
b. between thoracic vertebrae 11 and 12.
c. between lumbar vertebrae 1 and 2.
d. at lumbar vertebra 5.

7. There are _______________ pairs of spinal nerves.

8. Which of the following statements is/are true about the white matter in the spinal cord?

a. A cross section of the cord reveals a core of white matter surrounded by gray matter.
b. The white matter is composed of myelinated nerve fibers and makes up nerve pathways, called tracts.
c. The white matter carries sensory stimuli to the brain; the gray matter carries motor stimuli to the periphery.
d. The nerve fibers within spinal tracts arise from cell bodies located in the same part of the nervous system.

9. Pain impulses are carried from the area stimulated to the brain along the

a. fasciculus gracilis.
b. spinothalamic tracts.
c. fasciculus cuneatus.
d. spinocerebellar tract.

10. An individual with injury to the spinocerebellar tract is likely to experience

a. loss of a sense of touch.
b. uncoordinated movements.
c. involuntary muscle movements.
d. severely diminished pain perception.

11. An individual suffering from flaccid paralysis has most likely sustained damage to the _______________ tract.

a. spinocerebellar
b. corticospinal
c. ribrospinal
d. reticulospinal

12. The knee-jerk reflex is an example of a

a. reflex that controls involuntary behavior.
b. pathologic reflex.
c. withdrawal reflex.
d. monosynaptic reflex.

13. An individual who experiences a withdrawal reflex experiences pain at the same time the affected part is removed from the harmful stimulus.

a. True
b. False

14. Damage to the corticospinal tract in an adult may result in a/an _______________ reflex.

a. biceps-jerk
b. cremasteric
c. ankle-jerk
d. Babinski

15. Damage to upper motor neurons leads to _______________ paralysis.

a. spastic
b. flaccid
c. intermittent
d. alternating flaccid and spastic

16. Immediate, intensive treatment of spinal cord injuries is important to

a. begin regeneration of severed nerve fibers.
b. prevent extension of damage secondary to spinal shock.
c. both
d. neither

17. The cerebrum develops from a portion of the
   a. forebrain (prosencephalon).
   b. midbrain (mesencephalon).
   c. hindbrain (rhombencephalon).
18. The three major portions of the brain are the _______________, _______________, and _______________ _______________.
19. A neural tube defect in the lower posterior portion of the tube results in _______________ _______________.
20. The hemispheres of the cerebrum are connected by nerve fibers called the
   a. corpus callosum.
   b. falx cerebri.
   c. fissure of Rolando.
   d. tentorium.
21. The convolutions on the surface of the cerebrum are called
   a. sulci.
   b. fissures.
   c. gyri.
   d. ganglia.
22. Which of the following statements about the cerebral cortex is/are *true?*
   a. The cortex is the central white portion of the cerebrum.
   b. The cortex has sensory, motor, and association areas.
   c. The cortex is the outer gray area of the cerebrum.
   d. The cells in the right hemisphere of the cortex control the right side of the body.
23. Match the functions in the first column with the appropriate area of the brain in the second column.

____ 1. hearing	a. frontal lobes
____ 2. vision	b. parietal lobes
____ 3. recognition of printed words	c. temporal lobes
____ 4. control of voluntary muscles	d. occipital lobes
____ 5. pain	
____ 6. complex problem solving	

24. Damage to Broca s area in the cerebral cortex results in the inability to _______________.
25. Centers for higher intellectual functions, such as planning and complex problem solving, are located in the _______________ lobes.
26. Damage to the parietal lobes would impair an individual s ability to
   a. hear speech.
   b. understand speech.
   c. choose appropriate words in speaking.
   d. understand visual cues.
27. In most people, the _______________ hemisphere is dominant for verbal and computational skills.
28. Some investigators believe that intense, repetitive neuronal activity produces stable changes in nerve pathways to produce _______________ memory.
   a. short-term
   b. long-term
   c. collective
   d. unconscious
29. A brief interruption of blood flow to the brain produces a condition known as a _______________ _______________ _______________.
30. The function of basal ganglia is to
   a. inhibit emotional responses.
   b. inhibit motor functions.
   c. aid in temperature control.
   d. integrate hormonal function.
31. Cerebrospinal fluid is produced by the _______________ _______________ which is located in the cerebral _______________.
32. The neurotransmitter implicated in the production of the symptoms of Parkinson's disease is
   a. serotonin.
   b. dopamine.
   c. GABA.
   d. norepinephrine.

33. The major functions of cerebrospinal fluid include all of the following EXCEPT to

a. supply nutrients to parts of the brain.
b. prevent injury by acting as a shock absorber.
c. maintain the electrolyte balance of the cerebrum.
d. maintain temperature equilibrium in the brain stem.

34. An infant who has increasing pressure in the ventricles is said to have _______________.

35. The thalamus, hypothalamus, optic chiasma, and pituitary gland are parts of the brain located in the

a. midbrain.
b. pons.
c. medulla oblongata.
d. diencephalon.

36. The relay station that receives all sensory impulses except smell is the

a. pons.
b. medulla.
c. basal ganglia.
d. thalamus.

37. The part of the brain responsible for the regulation of temperature and heart rate, control of hunger, and regulation of fluid and electrolytes is the

a. thalamus.
b. hypothalamus.
c. medulla oblongata.
d. pons.

38. The _______________ _______________ produces emotional reactions of fear, anger, and pleasure.

39. The red nucleus of the midbrain is the center for

a. color vision.
b. eye reflexes.
c. postural reflexes.
d. temperature control.

40. The area of the brain that contains control centers for vital visceral functions is the _______________ _______________.

41. Consciousness is dependent on stimulation of the _______________ _______________.

42. REM sleep is also called _______________ sleep.

43. With the eyes closed, a person can accurately describe the positions of the various body parts. Which of the following structures serves in this function?

a. proprioceptors
b. pons
c. frontal lobe of the cerebrum
d. cerebellum

44. An individual who sustains damage to the cerebellum is likely to exhibit

a. tremors.
b. garbled speech.
c. bizarre thought patterns.
d. a loss of peripheral vision.

45. The peripheral nervous system has two divisions, the _______________ nervous system and the _______________ nervous system.

46. The nerve fibers that carry motor impulses to smooth muscle structures causing them to contract and to glands causing them to secrete are

a. general somatic afferent fibers.
b. general somatic efferent fibers.
c. general visceral efferent fibers.
d. general somatic afferent fibers.

47. There are _______________ pairs of cranial nerves; all but one of these arise from the _______________ _______________.

48. The cranial nerve that raises the eyelid and focuses the lens of the eye is the

a. optic nerve.
b. oculomotor nerve.
c. abducens nerve.
d. facial nerve.

49. In shrugging the shoulders, the sternocleidomastoid and trapezius muscles are stimulated by

a. the vagus nerve.
b. the trigeminal nerve.
c. the accessory nerve.
d. the hypoglossal nerve.

50. The anterior branches of the lower four cervical nerves and the first thoracic nerve give rise to the _______________ plexus.

51. Which of the following nerves arises from the lumbosacral plexus?

a. musculocutaneous nerve
b. femoral nerve
c. common peroneal nerve
d. medial nerve

52. The part of the nervous system that functions without conscious control is the ______________ nervous system.

53. Nerves of the sympathetic division leave the spinal cord with spinal nerves in the ______________ and ______________.

54. Nerves of the parasympathetic division leave the central nervous system within ______________ nerves and ______________ nerves.

55. Match the parts in the first column with the appropriate division in the second column.

____ 1. adrenergic fibers
____ 2. cholinergic fibers
____ 3. norepinephrine
____ 4. acetylcholine

a. sympathetic division
b. parasympathetic division

56. Which of the following are responses to stimulation by the sympathetic nervous system?

a. increased heart rate
b. increased blood glucose concentration
c. increased peristalsis
d. increased salivation

57. Which of the following are responses to stimulation of the parasympathetic nervous system?

a. dilation of the bronchioles
b. dilation of the coronary arteries
c. contraction of the gallbladder
d. contraction of the muscles of the urinary bladder

58. Control of the autonomic nervous system is in the ______________ and ______________ ______________; integration of autonomic function is in the ______________.

STUDY ACTIVITIES

I. Definition of Key Terms

Define the following key terms used in this chapter.

adrenergic

autonomic nervous system

brain stem

cerebral cortex

cerebral hemisphere

cerebrospinal fluid

cerebrum

cholinergic

choroid plexus

diencephalon

hypothalamus

medulla oblongata

midbrain

parasympathetic

postganglionic

preganglionic

reticular formation

sympathetic

thalamus

ventricle

II. Introduction and Meninges (pp. 375 77)

A. How do injuries to the brain, such as those sustained by Phineas Gage and Karen Ann Quinlan, increase our knowledge of how the brain functions?

B. What are the bony coverings of the central nervous system?

C. Fill in the following chart.

The meninges

Layer	Location	Structure and special features	Function
Dura mater			
Arachnoid mater			
Pia mater			

D. How is the structure of the dura mater related to the development of a subdural hematoma?

E. Which of the meninges are usually inflamed in meningitis?

III. Spinal Cord (pp. 377 82)

A. Answer the following concerning the structure of the spinal cord.

1. The superior boundary of the spinal cord is _______________. The inferior boundary of the spinal cord is _______________.

2. How many pairs of spinal nerves are there?

3. Label the accompanying drawing of a cross section of the spinal cord, showing the anterior median fissure, posterior median sulcus, white matter, gray matter, posterior horn, lateral horn, anterior horn, gray commissure, central canal, posterior funiculus, anterior funiculus, lateral funiculus, dorsal root of spinal nerve, ventral root of spinal nerve, dorsal root ganglion, spinal nerve.

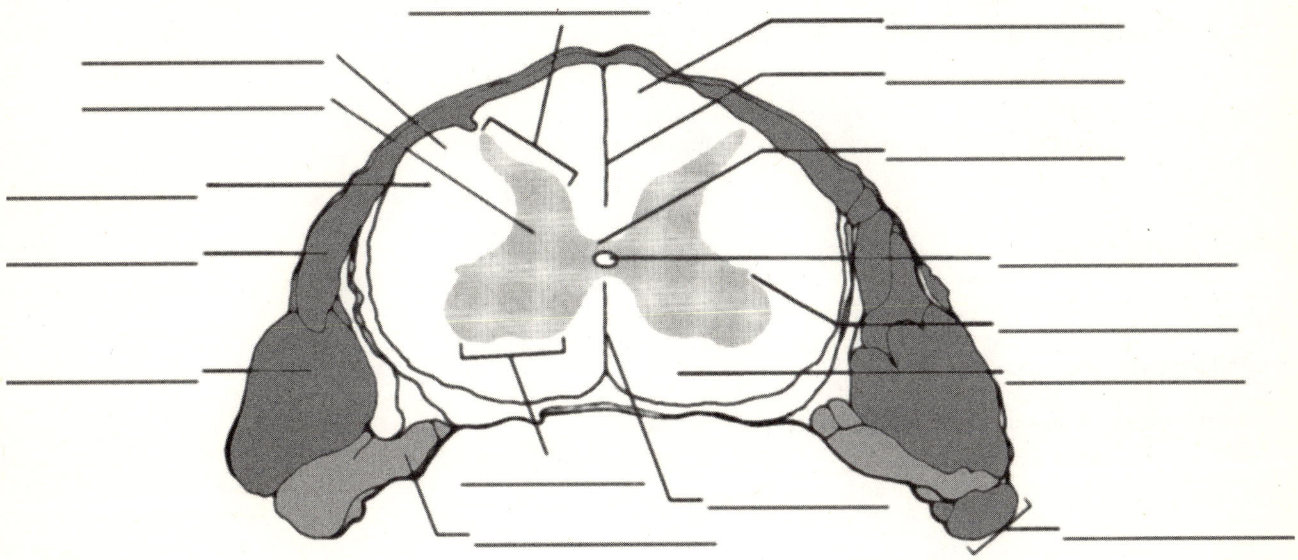

B. Fill in the following charts.

Ascending tracts of the spinal cord

Tract	Location	Function
Fasciculus cuneatus		
Fasciculus gracilis		
Spinothalamic		
Spinocerebellar		

Descending tracts of the spinal cord		
Tract	**Location**	**Function**
Corticospinal		
Reticulospinal		
Rubrospinal		

C. Answer the following questions about spinal cord pathology.

1. What is the effect of injury to ascending tracts? To descending tracts?

2. Describe spinal shock, flaccid paralysis, and spastic paralysis.

IV. Reflex Arcs (pp. 383–87)

Answer the following questions about reflex arcs and reflex behavior.

1. What is a reflex arc?

2. What is reflex behavior?

3. Label the parts of the reflex shown in the following drawing: dendrite of sensory neuron, cell body of sensory neuron, axon of sensory neuron, dendrite of motor neuron, cell body of motor neuron, axon of motor neuron, spinal cord, effector-quadriceps femoris muscle group, receptor ends of sensory neuron, tibia, patella, femur.

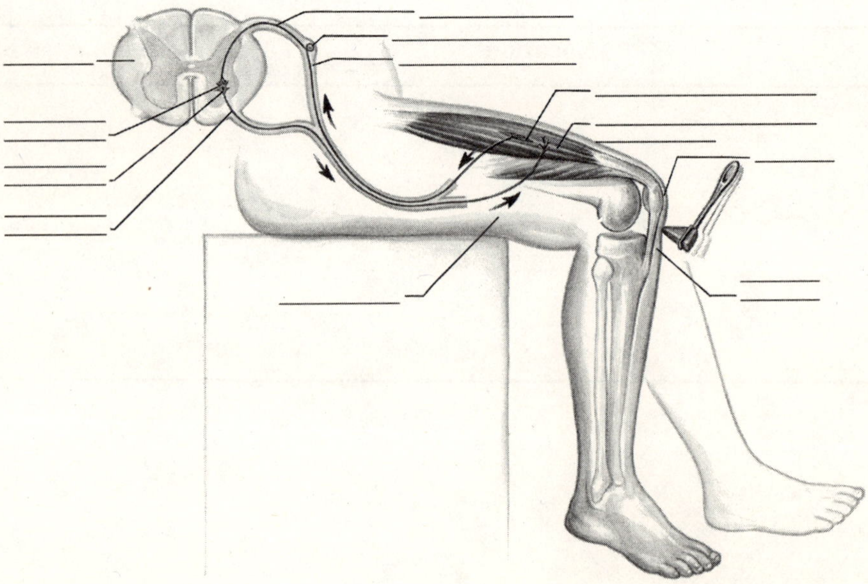

4. Describe the following types of reflexes.

   knee-jerk reflex

   withdrawal reflex

5. What information is yielded by testing normal reflexes? List and locate the common reflexes elicited during a neurological examination.

6. What structure is damaged in amyotrophic lateral sclerosis? How is the patient affected?

V. **Brain** (pp. 387 406)

A. Answer the following questions about the brain and its development.

1. The brain is composed of billions of _______________ neurons and nerve fibers.
2. What is the function of the brain?

3. Describe the development of the brain.

4. How is the structure of the brain related to this developmental process?

5. How are anencephaly and spina bifida related to brain development?

B. Answer the following concerning the structure of the cerebrum.

1. List the lobes of the cerebral hemispheres and describe the function of each.

2. The bridge that connects the two hemispheres is the _______________ _______________.
3. The ridges of the hemispheres are _______________ _______________.
4. A shallow groove is a _______________; a deeper groove is a _______________.
5. The outer layer of the cerebrum is the _______________.
6. The inner layer of the cerebrum is composed of _______________ _______________.
7. Masses of gray matter deep within the cerebrum that inhibit motor activity are the _______________ _______________.
8. Describe the disorder lissencephaly (smooth brain).

9. What noninvasive diagnostic technique enables researchers to study the functions of various parts of the living brain?

C. Fill in the following chart.

Functional areas of the cerebrum

	Location	Function
Motor		
Sensory		
Association		

D. What is meant by hemisphere dominance? Describe its impact on cerebral function.

E. Describe the processes involved in short-term memory and long-term memory.

F. Describe how a concussion occurs and list its symptoms.

G. Compare the events and results of a cerebrovascular accident and a transient ischemic attack (TIA).

H. Contrast the result of damage to the interpretative area in a child and in an adult.

I. Answer the following about the basal ganglia.

1. Basal ganglia are located within the _______________ _______________.
2. The substance produced by the ganglia is _______________.
3. Describe the symptoms and treatment of Parkinson s disease.

J. Answer the following concerning the ventricles and cerebrospinal fluid.

1. Describe the location of the four ventricles.
2. Where is cerebrospinal fluid secreted? What is its composition?
3. Describe the circulation of cerebrospinal fluid.
4. What are the functions of cerebrospinal fluid?
5. Why is the lumbar puncture done between the third and fourth or between the fourth and fifth lumbar vertebrae?

6. What is the clinical significance of the pressure of cerebrospinal fluid?

K. Answer these questions concerning the brain stem.

1. Where is the brain stem? What are its component structures?

2. Locate the diencephalon and describe its structure. Then fill in the following chart.

Structure of the diencephalon

Structure	Location	Function
Optic chiasma		
Pituitary gland		
Thalamus		
Hypothalamus		

3. What structures make up the limbic system? What is the function of this system?

4. Fill in the following chart.

Functions of the midbrain

Structure	Location	Function
Cerebral peduncles		
Corpora quadrigemina		
Red nucleus		

5. Where is the pons located? What is its function?

6. Where is the medulla oblongata located? What vital activities does the medulla control?

7. Describe the location, structure, and function of the reticular formation.

8. Describe normal (slow-wave) sleep and paradoxical sleep.

9. List the symptoms of the following sleep disorders:
   Fatal familial insomnia

   Insomnia

   Narcolepsy

   Obstructive sleep apnea

   Parasomnias

   REM-sleep behavior disorder

   Restless-leg syndrome

   Sleep paralysis

10. Compare coma and persistive vegetative states.

L. What is the function of the cerebellum?

M. Answer the following questions about electroencephalograms.

1. Describe the four types of brain waves recorded by an electroencephalogram.

2. How is this information used clinically?

VI. Peripheral Nervous System (pp. 407 18)

A. Answer these questions concerning the parts of the peripheral nervous system.

1. What are the parts of the peripheral nervous system?

2. What is the function of each of the following peripheral nerve fibers?
   general somatic efferent fibers

general visceral efferent fibers

general somatic afferent fibers

general visceral afferent fibers

special visceral efferent fibers

special visceral afferent fibers

special somatic afferent fibers

3. What is the function of the somatic nervous system? The autonomic nervous system?

B. Try the following concerning the cranial nerves.

1. An easy way to memorize the names of the cranial nerves is to use the following sentence:

   On old Olympus towering tops a Finn and German viewed some hops.
   I II III IV V VI VII VIII IX X XI XII

   You need to be able to identify the cranial nerves by both name and number.

2. Using yourself or a partner, demonstrate how you might test each of the cranial nerves.

3. Fill in the following chart.

Cranial nerve function

Cranial nerve		**Sensory, motor, or mixed**	**Function**
I	Olfactory		
II	Optic		
III	Oculomotor		
IV	Trochlear		
V	Trigeminal		
VI	Abducens		
VII	Facial		
VIII	Vestibulocochlear (acoustic)		

IX	Glossopharyngeal
X	Vagus
XI	Accessory (spinal accessory)
XII	Hypoglossal

4. What are the consequences of injury to any of the cranial nerves?

C. Answer these questions about spinal nerves.

1. How are the spinal nerves identified?

2. What are the structure and function of the dorsal root? Of the ventral root?

3. An area of skin in which a group of sensory nerves lead to a particular dorsal root is called a _______________.

4. What structures are innervated by the meningeal branch, the posterior branch, and the anterior branch of a spinal nerve?

5. What nerves have a visceral branch?

D. Answer the following concerning spinal nerve plexuses.

1. What is a plexus?

2. Fill in the following chart.

Spinal nerve plexuses

	Nerves involved	Structures innervated
Cervical plexus		
Brachial plexus		
Lumbosacral plexus		

E. Answer the following questions about injury to spinal nerves.

1. What is a whiplash injury?

2. Damage to the phrenic nerve results in paralysis of the _______________.
3. How does prolonged adduction of the arm affect the brachial plexus?

4. Describe sciatica.

VII. Autonomic Nervous System (pp. 418 26)

A. What structures make up the autonomic nervous system, and what is the function of this system?

B. How are the nerve pathways of the autonomic division different from those of the somatic division?

C. Identify the structural and functional differences between the sympathetic and parasympathetic divisions of the autonomic nervous system. Be sure to include differences in neurotransmitters.

D. Describe the interaction between acetylcholine and muscarinic and nicotinic cholinergic receptors.

E. How does the action of norepinephrine and epinephrine depend on alpha and beta receptors?

F. Describe the mechanisms of control of the autonomic nervous system.

VIII. Clinical Focus Question

Many nursing homes have extensive activity programs for their elderly residents. Based on your knowledge of the function of the central nervous system, explain the rationale for such programs in the treatment and rehabilitation of the elderly.

When you have completed the study activities to your satisfaction, retake the mastery test. If there are still some areas you do not understand, repeat the appropriate study activities.

CHAPTER 12
SOMATIC AND SPECIAL SENSES

OVERVIEW

This chapter deals with specialized parts of the nervous system that allow the body to assess and adjust to the external environment. It describes the types of receptors (objective 1) and their locations, structures, and functions in maintaining homeostasis (objectives 2–14).

An understanding of these senses is necessary to knowing how the nervous system receives input and responds to support life.

CHAPTER OBJECTIVES

After you have studied this chapter, you should be able to:

1. Name five kinds of receptors and explain the function of each.
2. Explain how receptors stimulate sensory impulses.
3. Explain how a sensation is produced.
4. Distinguish between somatic and special senses.
5. Describe the receptors associated with the senses of touch and pressure, temperature, and pain.
6. Describe how the sense of pain is produced.
7. Explain the importance of stretch receptors in muscles and tendons.
8. Explain the relationship between the senses of smell and taste.
9. Name the parts of the ear and explain the function of each part.
10. Distinguish between static and dynamic equilibrium.
11. Name the parts of the eye and explain the function of each part.
12. Explain how the eye refracts light.
13. Explain how depth and distance are perceived.
14. Describe the visual nerve pathway.

FOCUS QUESTION

When you began this chapter at 3:00 P.M., it was 32°F outside, but the sun was pouring into the room. It is now after 5:00 P.M. As you reach to turn on the light, you notice the room has become chilly, so you get a sweater. You smell the supper your roommate is preparing, and you realize that you are hungry. How have your special senses functioned to process and act on this sensory information?

MASTERY TEST

Now take the mastery test. Do not guess. Some questions may have more than one correct answer. As soon as you complete the test, correct it. Note your successes and failures so that you can read the chapter to meet your learning needs.

1. A colorblind person will perceive red as
   a. yellow.
   b. green.
   c. blue.
   d. pink.
2. Sensory receptors are sensitive to stimulation by
   a. changes in the concentration of chemicals.
   b. temperature changes.
   c. tissue damage.
   d. mechanical forces.
   e. changes in the intensity of light.

3. Sensory receptors for all of the following adapt to repeated stimulation by sending fewer and fewer impulses, except those for

a. heat.
b. light.
c. pain.
d. touch.

4. The senses of touch, pressure, temperature, and pain are called _______________ senses.

5. Match the sense in the first column with the appropriate receptor from the second column.

____ 1. touch and pressure
____ 2. light touch and texture
____ 3. heat
____ 4. cold
____ 5. deep pressure
____ 6. tension of muscle

a. Golgi tendon organs
b. free nerve endings
c. Pacinian corpuscles
d. Meissner's corpuscles

6. Pain receptors are sensitive to all of the following EXCEPT

a. chemicals such as histamine, kinins, hydrogen ions, and others.
b. electrical stimulation.
c. extremes of pressure.
d. extremes of heat and cold.

7. Heat relieves some kinds of pain by

a. increasing the metabolism in injured cells.
b. increasing blood flow to painful tissue.
c. decreasing the membrane permeability of sensory nerve fibers.
d. overriding pain sensation with heat sensation.

8. Pain perceived as a dull, aching sensation that is difficult to locate precisely is

a. chronic pain
b. referred pain
c. functional pain
d. visceral pain

9. Awareness of pain begins when pain impulses reach the

a. spinal cord.
b. medulla.
c. thalamus.
d. cerebral cortex.

10. The area of the body responsible for locating sources of pain or exerting control over emotional responses is the

a. limbic system.
b. thalamus.
c. reticular formation.
d. cerebral cortex.

11. Which of the following events will elicit pain from visceral organs?

a. spasm of smooth muscle
b. cutting into the viscera
c. stretching of a visceral organ
d. burning, as in electrocautery

12. Pain from the heart is likely to be experienced in the left shoulder. This is an example of _______________ pain.

13. The impulses that create a pain sensation that seems sharp and localized to a specific area, and that seems to originate in the skin and to disappear when the stimulus is removed, are likely to be transmitted on (acute/chronic) pain fibers.

14. With the exception of impulses arising from tissues of the head, pain impulses are carried on _______________ nerves.

15. A group of neuropeptides that have pain-suppressing activity and are released by the pituitary gland and the hypothalamus are _______________.

16. The senses of vision, taste, smell, hearing, and equilibrium are called _______________ senses.

17. The receptors for taste and smell are examples of

a. mechanical receptors.
b. chemoreceptors.
c. thermoreceptors.

18. Olfactory receptors are located in

a. the nasopharynx.
b. the inferior nasal conchae.
c. the superior nasal conchae.
d. the lateral wall of the nostril.

19. Impulses that stimulate the olfactory receptors are transmitted along the _______________ _______________.

20. A condition in which a sensation involving one sense is perceived in terms of another is ____________________.

21. The sensitive part of a taste bud is the taste
    a. cell.
    b. pore.
    c. hair.

22. Saliva enhances the taste of food by
    a. increasing the motility of taste receptors.
    b. dissolving the chemicals that cause taste.
    c. releasing taste factors by partially digesting food.

23. The four primary taste sensations are ______________, ______________, ______________, and ______________.

24. Sense of taste is strongly related to which of the other special senses? ______________

25. In addition to the sense of hearing, the ear also functions in the sense of ______________.

26. The waxy substance secreted by glands in the external ear is ______________.

27. The functions of the small bones of the middle ear are to
    a. provide a framework for the tympanic membrane.
    b. protect the structures of the inner ear.
    c. transmit vibrations from the external ear to the inner ear.
    d. increase the force of vibrations transmitted to the inner ear.

28. The skeletal muscles in the middle ear function to
    a. maintain tension in the eardrum.
    b. move the external ear.
    c. equalize the pressure on both sides of the eardrum.
    d. protect the inner ear from damage from loud noise.

29. The function of the eustachian tube is to
    a. prevent infection.
    b. intensify sound.
    c. equalize pressure.
    d. modify pitch.

30. Ear infections are more common in children than in adults because
    a. children have immature immune systems.
    b. blood supply to the middle ear is less in children than in adults.
    c. the eustachian tube is shorter in children than in adults.
    d. young children are likely to suck their thumbs.

31. The inner ear consists of two complex structures called the ______________ ______________ and the ______________ ______________.

32. Sound is transmitted in the inner ear via a fluid called ______________.

33. Hearing receptors are located in the
    a. organ of Corti.
    b. scala vestibuli.
    c. scala tympani.
    d. round window.

34. Impulses from hearing receptors are transmitted via the
    a. abducens nerve.
    b. facial nerve.
    c. cochlear branch of the vestibulocochlear nerve.
    d. trigeminal nerve.

35. Otosclerosis is classified as ______________ deafness.

36. Prolonged exposure to noise, tumors, and some antibiotics are causes of ______________ deafness.

37. A cochlear implant may be used to treat ______________ deafness.

38. The organs concerned with static equilibrium are located within the ______________.

39. The hair cells of the crista ampullaris are stimulated by
    a. bending the head forward or backward.
    b. rapid turns of the head or body.
    c. changes in the position of the body relative to the ground.
    d. changes in the position of skeletal muscles.

40. The muscle that raises the eyelid is the

a. orbicularis oculi.
b. superior rectus.
c. levator palpebrae superioris.
d. ciliary muscle.

41. The lacrimal gland is located in the _______________ of the orbit.

a. superior lateral wall
b. superior medial wall
c. inferior lateral wall
d. inferior medial wall

42. The conjunctiva covers the anterior surface of the eyeball, except for the _______________.

43. The superior rectus muscle rotates the eye

a. upward and toward the midline.
b. toward the midline.
c. away from the midline.
d. upward and away from the midline.

44. The orbicularis oculi is innervated by the

a. oculomotor nerve.
b. trochlear nerve.
c. abducens nerve.
d. facial nerve.

45. The transparency of the cornea is due to

a. the nature of the cytoplasm in the cells of the cornea.
b. the small number of cells and the lack of blood vessels.
c. the lack of nuclei with these cells.
d. keratinization of cells in the cornea.

46. In the posterior wall of the eyeball, the sclera is pierced by the _______________.

47. The first successful human organ transplant was transplantation of the _______________.

48. The anterior portion of the middle tunic or vascular tunic of the eye contains the

a. choroid coat.
b. ciliary body.
c. iris.
d. cornea.

49. The shape of the lens changes as the eye focuses on a close object in a process known as

a. accommodation.
b. refraction.
c. reflection.
d. strabismus.

50. An increasing opacity in the lens of the eye is a(n) _______________.

51. The anterior chamber of the eye extends from the _______________ to the iris.

52. The aqueous humor leaves the anterior chamber via the

a. pupil.
b. canal of Schlemm.
c. ciliary body.
d. lymphatic system.

53. The part of the eye that controls the amount of light entering it is the _______________.

54. The color of the eye is determined by the amount and distribution of _______________ in the iris.

55. The inner tunic of the eye contains the receptor cells of sight and is called the _______________.

56. The region associated with the sharpest vision is the

a. macula lutea.
b. fovea centralis.
c. optic disk.
d. choroid coat.

57. The largest compartment of the eye, which is bounded by the lens, ciliary body, and retina, is filled with _______________ _______________.

58. The bending of light waves as they pass at an oblique angle from a medium of one optical density to a medium of another optical density is called _______________.

59. The lens loses elasticity with aging, causing a condition called _______________.

60. There are two types of visual receptors: one has long, thin projections that are called _______________; the other has short, blunt projections that are called _______________.

61. Match the type of vision in the first column with the proper receptor from the second column.

____ 1.	vision in relatively dim light	a.	rods
____ 2.	color vision	b.	cones
____ 3.	general outlines		
____ 4.	sharp images		

62. The light-sensitive pigment in rods is ______________. In the presence of light, this pigment decomposes to form ______________ and ______________.

63. The pigments found in cones are ______________, ______________, and ______________.

64. The absence of cone pigments leads to ______________, ______________.

65. If the visual cortex is injured, the individual may develop (complete/partial) blindness in (one eye/both eyes).

STUDY ACTIVITIES

I. Definition of Key Terms

Define the following key terms used in this chapter.

accommodation

ampulla

auditory

chemoreceptor

cochlea

cornea

dynamic equilibrium

labyrinth

macula

mechanoreceptor

olfactory

optic

photoreceptor

projection

proprioceptor

referred pain

refraction

retina

rhodopsin

sclera

sensory adaptation

static equilibrium

thermoreceptor

II. Introduction and Receptors and Sensations (pp. 432–34)

A. How is a sensation or feeling produced?

B. List five groups of sensory receptors and identify the sensations with which they are associated.

C. Describe the transmission of sensory impulses.

D. A feeling that occurs when sensory impulses are interpreted by the brain is a _______________.

E. The process that allows an individual to locate the region of stimulation is called _______________.

F. The process that makes a receptor ignore a continuous stimulus unless the strength of that stimulus increased is _______________ _______________.

III. Somatic Senses (pp. 434–39)

A. Fill in the following chart.

Cutaneous receptors

Type	Function	Sensation
Free nerve endings (mechanoreceptors)		
Meissner s corpuscles (mechanoreceptors)		
Free nerve endings (thermoreceptors–heat)		
Free nerve endings (thermoreceptors–cold)		
Free nerve endings (pain receptors)		

B. Answer the following concerning pain receptors.

1. In what way do pain receptors differ from the other somatic receptors?

2. Tissue damage is thought to stimulate pain receptors by the release of _______________.

3. Describe the "vicious cycle" that is set up when a skeletal muscle contracts in response to pain.

4. Compare the treatments for cancer pain and chronic pain.

5. What events trigger visceral pain?

6. What is referred pain?

7. Compare the characteristics of acute pain fibers and chronic pain fibers.

8. How are pain impulses regulated?

9. Neurotransmitters secreted by the spinal cord and thalamus that inhibit pain impulses are _______________.

10. A pain suppressant secreted by the pituitary gland is _______________.

C. Discuss the interaction of stretch receptors in the muscles (muscle spindles) and the tendons (Golgi tendon organs).

IV. Sense of Smell (pp. 440–42)

A. The sense of smell supplements the sense of _______________.

B. On the accompanying illustration, label the olfactory tract, olfactory bulb, cribriform plate, nasal cavity, olfactory area of the nasal cavity, superior nasal concha.

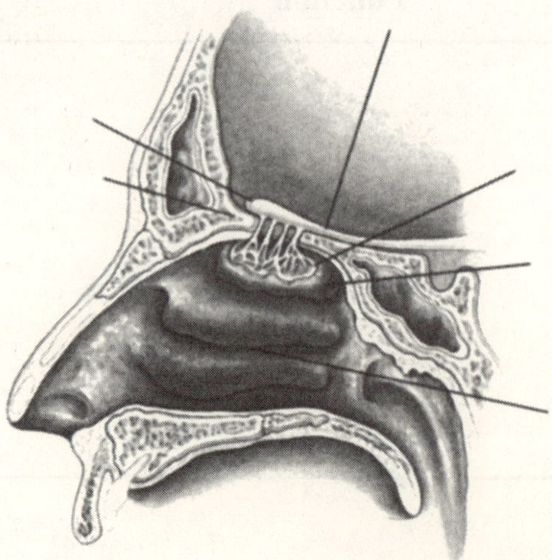

C. How do odors stimulate olfactory receptors?

D. What is synesthesia?

E. Describe the nerve pathways for the sense of smell.

F. Why does the sense of smell diminish in acuity with increasing age?

V. **Sense of Taste** (pp. 442–46)

A. Describe the structure of taste receptors.

B. How does saliva contribute to the perception of taste?

C. Identify the primary taste sensation associated with each darkened area in the accompanying illustration.

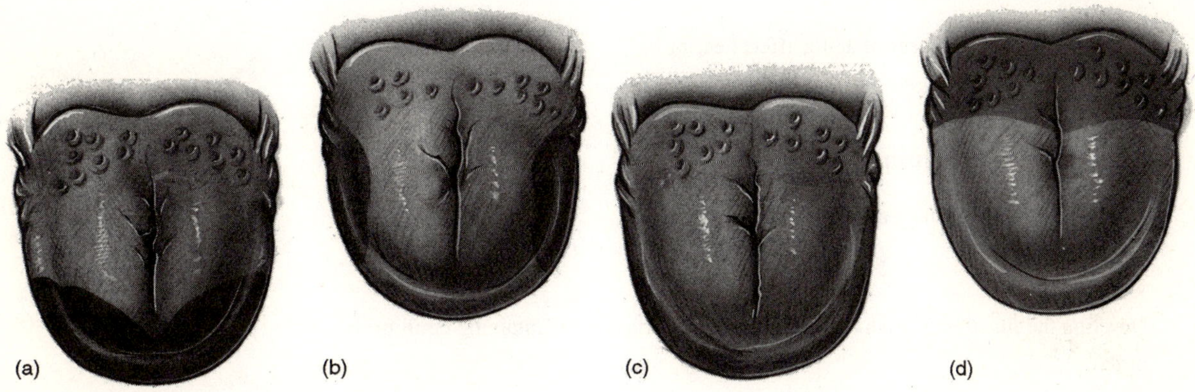

D. Describe the nerve pathways for the sensation of taste.

E. What factors can influence or distort an individual's senses of smell or taste?

VI. **Sense of Hearing** (pp. 446–51)

A. Describe the function of the external ear.

B. Describe the vibration conduction pathway of the ear from the meatus to the temporal lobe of the cerebrum.

C. Why does it help to chew gum while descending in an airplane?

D. How does the structure of the auditory tube and middle ear predispose the middle ear to infection?

E. Why can't the tympanic reflex protect the hearing receptors from the effects of sudden, loud noises?

F. Describe the function of the inner ear.

G. Complete these statements concerning deafness.

1. An interference with the transmission of sound to the inner ear leads to _______________ deafness.
2. Damage to the nervous structures, which can be caused by loud sounds, leads to _______________ deafness.
3. Describe the Weber and Rinne tests.

4. What is a cochlear implant?

5. How does the process of aging affect hearing?

VII. Sense of Equilibrium (pp. 451–57)

A. Distinguish between static and dynamic equilibrium.

B. Describe the function of each of the following structures in maintaining equilibrium.

utricle

saccule

macula

semicircular canals

crista ampullaris

cerebellum

eyes

C. What is motion sickness?

VIII. Sense of Sight (pp. 457–75)

A. Answer the following concerning the visual accessory organs.

1. What structures are covered by the conjunctiva?

2. Describe the lacrimal apparatus. How does it protect the eye?

3. What is the secretion of the conjunctiva and what is its function?

4. Identify the function of the following muscles.

   orbicularis oculi

   levator palpebrae superioris

   superior rectus

   inferior rectus

   medial rectus

   lateral rectus

   superior oblique

   inferior oblique

5. Why is vision screening important for preschoolers?

B. Label these structures in the following illustration: cornea, lens, iris, suspensory ligaments, vitreous humor, aqueous humor, sclera, optic disk, optic nerve, fovea centralis, posterior cavity, choroid coat, pupil, retina, anterior chamber, posterior chamber, ciliary body. What is the function of each of the labeled structures?

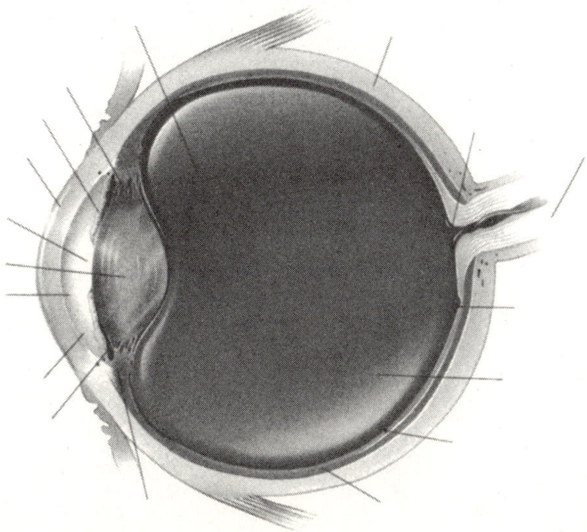

C. What characteristics of the cornea contribute to the ease with which it is transplanted?

D. Answer the following questions about diseases of the eye.

1. What is a cataract?

2. Describe the mechanisms and treatment of glaucoma.

3. What are floaters?

E. Answer these statements and questions concerning refraction of light.

1. Define refraction of light.

2. Which one of the following illustrates normal refraction? Identify the problems illustrated in the other two drawings.

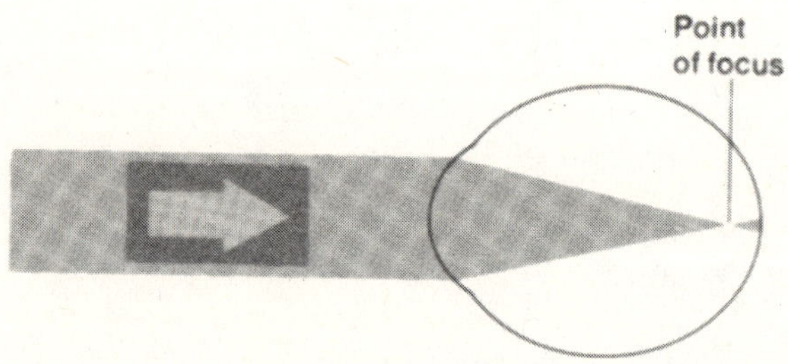

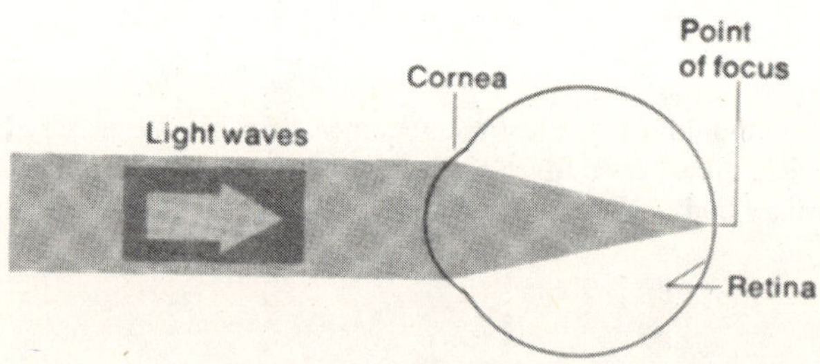

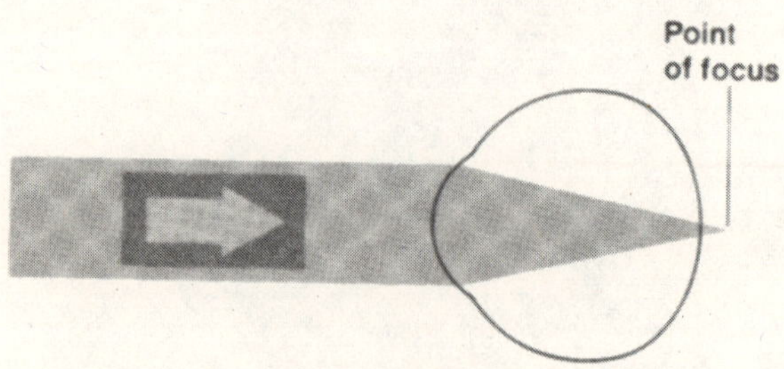

F. Answer the following concerning visual receptors.

1. Describe the functions of rods.

2. How are rods related to dark adaptation?

3. Describe the location and functions of cones.

4. What mechanism is used by cones to recognize color?

5. Why does vitamin A deficiency affect vision?

6. Describe colorblindness.

G. Answer the following concerning visual nerve pathways.

1. Describe the paths of the right and left optic tracts.

2. Describe the visual deficit that results from damage to either visual cortex.

IX. Clinical Focus Question

Imaging is a pain control technique in which patients are asked to visualize a place that is appealing to them, such as a seashore. They are then asked to concentrate on that image to relieve their pain. Based on your knowledge of the sense of pain, explain why this technique is effective.

When you have completed the study activities to your satisfaction, retake the mastery test and compare your performance with your initial attempt. If there are still areas you do not understand, repeat the appropriate study activities.

CHAPTER 13
ENDOCRINE SYSTEM

OVERVIEW

The endocrine system, like the nervous system, controls body activities to maintain a relatively constant internal environment. The methods used by these two systems are different. This chapter describes the difference between endocrine and exocrine glands, the location of the endocrine glands, and the hormones they secrete (objectives 1 and 6). It explains the nature of hormones, the substances that function as hormones, how hormones affect target tissues, how the secretion of hormones is controlled by a negative feedback system and the nervous system, the general function of each hormone, and the result of too little or too much of each hormone (objectives 2–5, 7, and 8). In addition, the text distinguishes between physical and psychological stress and describes how the endocrine system mediates the stress response (objectives 9 and 10).

A knowledge of the function of the endocrine system is basic to the understanding of how metabolic processes are regulated to meet the changing needs of the body.

CHAPTER OBJECTIVES

After you have studied this chapter, you should be able to:

1. Distinguish between endocrine and exocrine glands.
2. Describe how hormones can be classified according to their chemical composition.
3. Explain how steroid and nonsteroid hormones affect target cells.
4. Discuss how negative feedback mechanisms regulate hormonal secretions.
5. Explain how the nervous system controls hormonal secretions.
6. Name and describe the locations of the major endocrine glands, and list the hormones they secrete.
7. Describe the general functions of the various hormones.
8. Explain how the secretion of each hormone is regulated.
9. Distinguish between physical and psychological stress.
10. Describe the general stress response.

FOCUS QUESTION

How do the functions of the nervous system and the endocrine system differ, and how do they complement one another?

Mastery Test

Now take the mastery test. Do not guess. Some questions may have more than one correct answer. As soon as you complete the test, correct it. Note your successes and failures so that you can read the chapter to meet your learning needs.

1. An endocrine disease diagnosed by Egyptians as long ago as 1500 B.C. by the "melting down of limbs and flesh into urine" is _______________ _______________.
2. Glands that release their secretions into ducts that lead to the outside of the body are _______________ glands.
3. Glands that control the rate of chemical reactions, help transport substances through cell membranes, and help regulate fluid and electrolyte balance are _______________ glands.
4. Glands whose secretions affect only local, neighboring cells are _______________ glands.
5. Glands whose secretions affect only the secreting cell itself are ____________ glands.

6. Hormones belong to all of the following chemical families EXCEPT
   a. amines.
   b. polysaccharides.
   c. proteins.
   d. steroids.
7. Steroid hormones influence cells by
   a. altering the cell's metabolic processes.
   b. influencing the rate of cell reproduction.
   c. changing the nature of cellular protein.
   d. causing special proteins to be synthesized.
8. Cellular responses to nonsteroid hormones include
   a. alteration of cell membrane permeability.
   b. activating enzymes.
   c. promoting the synthesis of certain proteins.
   d. changing the activity of metabolic pathways.
   e. starting the secretion of hormones.
9. Which of the following substances is a common second messenger mediating the action of nonsteroid hormones?
   a. adenosine diphosphate
   b. adenosine triphosphate
   c. cyclic adenosine monophosphate
   d. G protein
10. The physiologic action of a hormone is determined by
   a. the conditions under which it is secreted.
   b. its chemical composition.
   c. its target cells.
   d. the amount of hormone secreted.
11. Steroid hormones are
   a. lipids.
   b. proteins.
   c. glycoproteins.
   d. carbohydrates.
12. Norepinephrine and epinephrine are examples of
   a. steroid hormones.
   b. amine hormones.
   c. peptide hormones.
   d. protein hormones.
13. An example of a protein hormone is those secreted by the
   a. pituitary gland.
   b. thyroid gland.
   c. adrenal gland.
   d. parathyroid gland.
14. Prostaglandins are lipids that are synthesized from ________________ ____________.
15. Athletes may abuse ______________ ________ to increase muscle size and improve athletic performance.
16. Prostaglandins have hormonelike effects and are thought to act by regulating
   a. the rate of mitosis.
   b. the production of cyclic AMP.
   c. cellular oxidation
   d. the utilization of glucose.
17. The characteristics of the negative feedback systems that regulate hormone secretion include
   a. activation by imbalance.
   b. exertion of an inhibitory effect on the gland.
   c. exertion of a stimulating effect on the gland.
   d. a tendency for levels of hormone to fluctuate.
18. The part of the brain most closely related to endocrine function is the _______________.
19. The hormones secreted by the anterior lobe of the pituitary gland include
   a. thyroid-stimulating hormones.
   b. luteinizing hormone.
   c. antidiuretic hormone.
   d. oxytocin.
20. Nerve impulses from the hypothalamus stimulate the _______________ lobe of the pituitary gland.
21. Which of the following are actions of pituitary growth hormone?
   a. enhance the movement of amino acids through the cell membrane
   b. increase the utilization of glucose by cells
   c. increase the utilization of fats by cells
   d. enhance the movement of potassium across the cell membrane

22. Which of the following conditions is/are likely to occur when the secretion of growth hormone is low during childhood?
    a. mental retardation
    b. short stature; well proportioned appearance
    c. small, short body; large head
    d. failure to develop secondary sex characteristics

23. An adult who suffers from the oversecretion of growth hormone is said to have ______________.

24. The pituitary hormone that stimulates and maintains milk production following childbirth is ______________.

25. Thyrotropin secretion is regulated by
    a. circulating thyroid hormones.
    b. blood sugar levels.
    c. the osmolarity of blood.
    d. TRH secreted by the hypothalamus.

26. Which of the following does follicle-stimulating hormone produce?
    a. growth of egg follicles
    b. production of estrogen
    c. production of progesterone
    d. production of sperm cells

27. Which of the following pituitary hormones helps maintain fluid balance?
    a. oxytocin
    b. ACTH
    c. antidiuretic hormone
    d. vasopressin

28. Antidiuretic hormone increases blood pressure by stimulating the contraction of the smooth muscle in blood vessels by increasing calcium ion concentration.
    a. True
    b. False

29. The thyroid hormones that affect the metabolic rate are ______________ and ______________.

30. Which of the following are functions of thyroid hormones?
    a. control sodium levels
    b. decrease rate of energy release from carbohydrates
    c. increase protein synthesis
    d. accelerate growth in children

31. The element necessary for normal function of the thyroid gland is ______________.

32. The thyroid hormone that tends to keep calcium in the bone is ______________.

33. A thyroid dysfunction characterized by exophthalmos, weight loss, excessive perspiration, and emotional instability is
    a. simple goiter.
    b. myxedema.
    c. hyperthyroidism.
    d. thyroiditis.

34. Which of the following statements about parathormone is/are true?
    a. Parathormone enhances the absorption of phosphorus and calcium from the intestine.
    b. Parathormone stimulates the bone to release ionized calcium.
    c. Parathormone stimulates the kidney to conserve calcium.
    d. Parathormone secretion is stimulated by the hypothalamus.

35. Injury to or removal of parathyroid glands is likely to result in
    a. reduced osteoclastic activity.
    b. Cushing's disease.
    c. kidney stones.
    d. hypocalcemia.

36. The hormones of the adrenal medulla are ______________ and ______________.

37. The adrenal hormone aldosterone belongs to a category of cortical hormones called
    a. mineralocorticoids.
    b. glucocorticoids.
    c. sex hormones.

38. The most important action(s) of cortisol in helping the body overcome stress is/are
    a. inhibition of protein synthesis to increase the levels of circulating amino acids.
    b. increasing the release of fatty acids and decreasing the use of glucose.
    c. stimulation of gluconeogenesis.
    d. conservation of water.

39. Adrenal sex hormones are primarily (male/female).

40. Masculinization of women, elevated blood glucose, decreases in tissue protein, and sodium retention are associated with

a. Addison's disease.
b. hypersecretion of adrenal cortical hormone.
c. Cushing's disease.
d. hyposecretion of adrenal cortical hormone.

41. The endocrine portion of the pancreas is made up of cells called _______________ _______________ _______________.

42. The hormone that responds to a low blood sugar by stimulating the liver to convert glycogen to glucose is _______________.

43. The actions of insulin include

a. enhancing glucose absorption from the small intestine.
b. facilitating the transport of glucose across the cell membrane.
c. promoting the transport of amino acids out of the cell.
d. increasing the synthesis of fats.

44. Hypoinsulinism results in a disease called _______________ _______________.

45. The endocrine gland(s) that seem(s) to influence circadian rhythms is/are the

a. thymus.
b. pineal gland.
c. gonads.

46. Stressors stimulate which of the following endocrine glands?

a. islets of Langerhans
b. parathyroid glands
c. adrenal cortex
d. adrenal medulla

47. The only valid claim for the use of melatonin supplements is

a. insomnia.
b. autism.
c. seizure disorders.
d. sudden infant death syndrome.

48. Thymosin, the secretion of the thymus gland, affects the production of certain white blood cells known as _______________________.

49. A person experiencing emotional stress is (more/less) likely to develop an infection than an individual with a lower stress level.

50. The hormone secreted by the kidney is _______________________.

STUDY ACTIVITIES

I. Definition of Key Terms

Define the following key terms used in this chapter.

adenylate cyclase

adrenal cortex

adrenal medulla

anterior pituitary

cyclic AMP

hormone

kinase

metabolic rate

negative feedback

pancreas

parathyroid gland

pineal gland

posterior pituitary

prostaglandin

steroid

target cell

thymus gland

thyroid gland

II. Introduction and General Characteristics of the Endocrine System (p. 480)

A. Diabetes mellitus can be controlled by regular injections of

a. insulin.
b. glucose.
c. glucagon.
d. pancreatic enzymes.

B. List the differences between endocrine and exocrine glands.

C. Two additional patterns of secretion are ____________________ glands and ________________ glands.

D. As a group, endocrine glands regulate _______________ _______________.

III. Hormone Action (pp. 480–87)

A. The specific site of a hormone's action is called its _______________ _______________.

B. Compare the mechanisms of action of steroid and nonsteroid hormones.

C. How do hormones recognize the target cells upon which they exert their effects?

D. List the chemical composition of hormones, and give examples of each type.

E. Compare hormones and prostaglandins.

F. Describe how steroid hormones act on their target tissue. Be sure to include the role of RNA.

G. Why do some athletes abuse anabolic steroids?

H. Describe two mechanisms of the action of nonsteroid hormones.

I. What is the role of prostaglandins in the response of target tissue to hormonal stimulation?

J. How is the knowledge of prostaglandins used in current medical practice?

IV. Control of Hormonal Secretions (pp. 487–88)

A. Answer the following concerning the negative feedback system.

1. Describe a negative feedback system.

2. How does it work in the endocrine system?

B. How does the nervous system control hormone secretion?

V. Pituitary Gland (pp. 488–495)

A. On the following illustration, label the cerebral cortex, hypothalamus, pituitary stalk, posterior and anterior lobes of the pituitary gland, third ventricle, sphenoidal sinus, sphenoid bone, sella turcica, anterior cerebral artery, optic nerve, optic chiasma, oculomotor nerve, trochlear nerve, basilar artery.

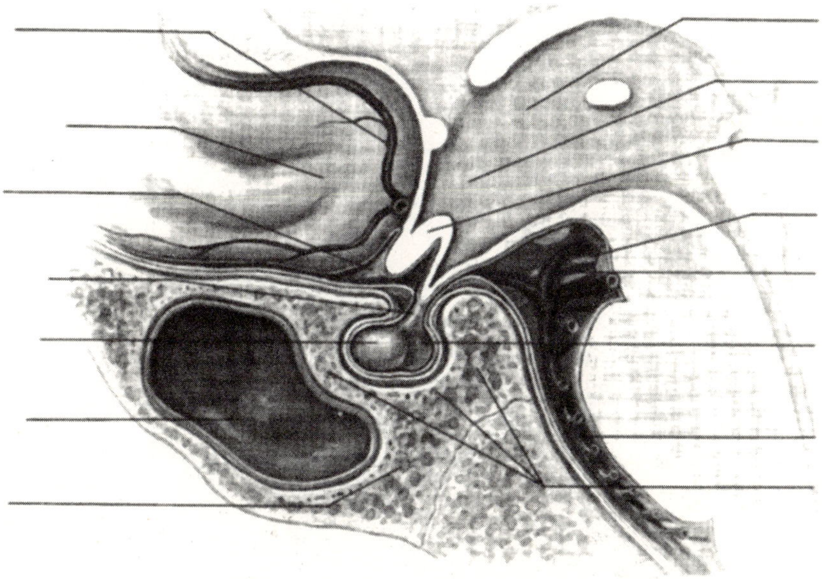

B. Fill in the following chart.

Hormones of the pituitary gland

Hormone	**Source of control**	**Actions (be specific)**
Anterior lobe		
Growth hormone		
Prolactin		
Thyroid-stimulating hormone		
Adrenocorticotropic hormone		
Follicle-stimulating hormone		
Luteinizing hormone		
Posterior lobe		
Antidiuretic hormone		
Oxytocin		

C. Answer the following questions about growth hormone.

1. Describe the "pulse secretion" of growth hormone.

2. What is the role of growth hormone and somatomedin in the growth of bone and cartilage?

D. Answer the following questions on pituitary dysfunction.

1. An insufficient amount of growth hormone in childhood is called _______________.
2. How is the condition in question 1 treated?

3. An oversecretion of growth hormone during childhood leads to a condition called _______________.
4. An oversecretion of growth hormone in an adult leads to a condition called _______________.
5. A disorder of ADH regulation that is manifested by increased urine production is _______________ _______________.

VI. Thyroid Gland (pp. 495–98)

A. On the following illustration, label the thyroid gland, isthmus, larynx, follicular cell, colloid, and extrafollicular cell.

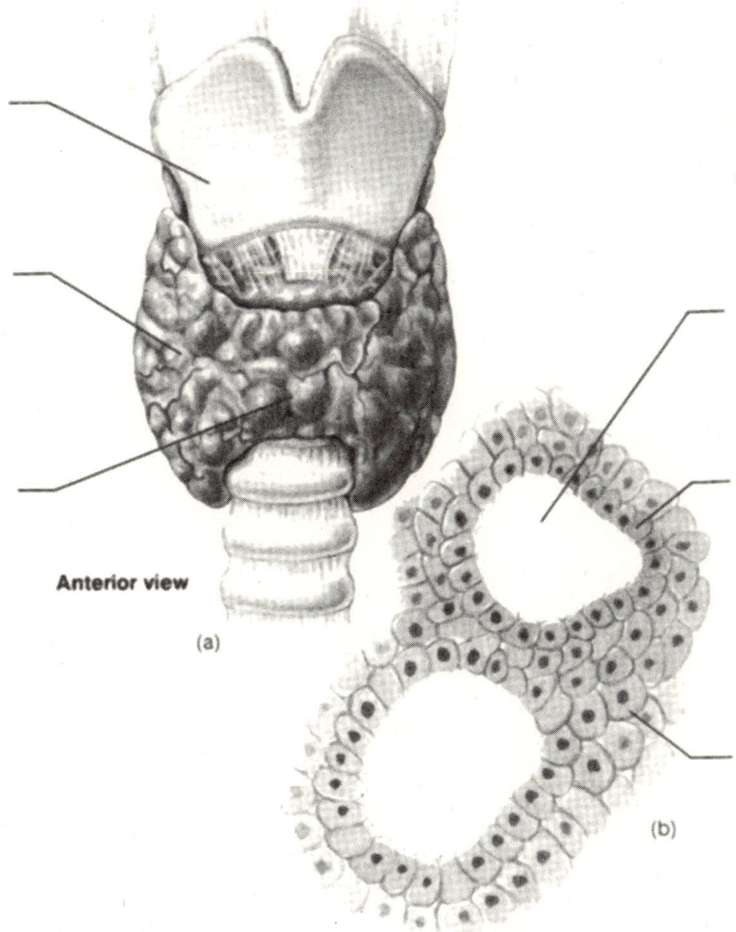

B. Answer the following questions concerning thyroid hormones and their functions.

1. What element is needed to synthesize thyroxine and triiodothyronine?

2. What is the function of thyroglobulin?

3. What are the functions of thyroxine and triiodothyronine?

4. What is the function of calcitonin?

C. Fill in the following chart.

Disorders of the thyroid

Disorder	Problems encountered	Treatment
Hyperthyroid		
Graves' disease		
Hyperthyroidism		
Hypothyroid		
Hashimoto's disease		
Hypothyroidism (infantile)		
Hypothyroidism (adult)		
Simple goiter		

VII. Parathyroid Glands (pp. 499–501)

A. Where are the parathyroid glands located?

B. Answer the following concerning parathyroid hormones.

1. Describe how parathormone affects blood levels of calcium and phosphorus. Include its effect on bone, the intestine, and the kidneys.

2. How does parathormone interact with calcitonin to maintain blood calcium levels?

C. Answer these questions concerning disorders of the parathyroid glands.

1. What happens when the parathyroid glands are overactive? How is this treated?
2. What causes hypoparathyroidism?
3. What are the symptoms of hypoparathyroidism? How is this treated?

D. Describe the effect of nonendocrine cancer cells.

VIII. Adrenal Glands (pp. 501–6)

A. Where are the adrenal glands located?

B. Answer the following concerning hormones of the adrenal medulla.

1. List the hormones secreted by the adrenal medulla.
2. What are the effects of these hormones?

C. Fill in the following chart.

Adrenocortical hormones

	Zone of the cortex	Stimulus for secretion	Effects of hormone
Mineralocorticoids (e.g., aldosterone)			
Glucocorticoids (e.g., cortisol)			
Sex hormones			

D. How do ACE inhibitors lower blood pressure?

E. What is the role of cortisol in the body's response to stress?

F. What are the symptoms and treatment of Addison's disease? Of Cushing's disease?

IX. Pancreas (pp. 507–10)

A. Where is the pancreas located?

B. Fill in the following chart.

Hormones of the pancreas

Hormone	Source of control	Effects of hormone
Glucagon		
Insulin		
Somatostatin		

C. Describe diabetes mellitus. Include the cause and the metabolic effects of this disease.

D. Compare insulin-dependent and noninsulin-dependent diabetes mellitus.

X. Other Endocrine Glands (pp. 510–11)

A. Where is the pineal gland located, and what is its function?

B. Where is the thymus gland located, and what is its function?

XI. Stress and Its Effects (pp. 511–12)

A. Define *stress* and list the factors that lead to stress.

B. Describe the body's response to stress. Be sure to include changes in hormone secretion and the positive aspects of the stress response.

XII. Clinical Focus Question

Based on your knowledge of the endocrine system, what symptoms might an individual who had a tumor of the pituitary gland experience?

When you have completed the study activities to your satisfaction, retake the mastery test and compare your performance with your initial attempt. If there are still areas you do not understand, repeat the appropriate study activities.

CHAPTER 14
BLOOD

OVERVIEW

This chapter deals with a major connective tissue—blood. It describe the general characteristics and major functions of blood (objective 1). This chapter identifies the various types of blood cells, and the components and functions of blood plasma (objectives 2 and 6). It explains blood cell counts, blood cell typing, red blood cell life cycle, control of red blood cell production, blood coagulation, and reaction to mixing of blood types (objectives 3–5 and 7–11).

Knowledge of the blood and its components expands the concept of how cells meet their need for oxygen. This knowledge also helps in understanding how the body recognizes and rejects foreign protein.

CHAPTER OBJECTIVES

After you have studied this chapter, you should be able to:

1. Describe the general characteristics of the blood and discuss its major functions.
2. Distinguish among the types of blood cells.
3. Explain how blood cell counts are made and how they are used.
4. Discuss the life cycle of a red blood cell.
5. Explain control of red blood cell production.
6. List the major components of blood plasma and describe the functions of each.
7. Define *hemostasis,* and explain the mechanisms that help to achieve it.
8. Review the major steps in blood coagulation.
9. Explain how coagulation can be prevented.
10. Explain the basis for blood typing.
11. Describe how blood reactions may occur between fetal and maternal tissues.

FOCUS QUESTION

How does the structure of the blood help to meet oxygenation needs, allow recognition and rejection of foreign protein, and control coagulation of the blood?

MASTERY TEST

Now take the mastery test. Do not guess. Some questions may have more than one correct answer. As soon as you complete the test, correct it. Note your successes and failures so that you can read the chapter to meet your learning needs.

1. Blood from the umbilical cord is useful for transplantation because it has large amounts of
   a. fetal hemoglobin.
   b. stem cells.
   c. neutrophils.
   d. platelets.
2. The percentage of formed elements in blood is called the _______________.
3. The intercellular material of blood is _______________.
4. Plasma represents ______% of a normal blood sample.
5. Cellular components of the immune system and formed elements of blood originate from a common stem cell known as a hematocyte.
   a. True
   b. False

6. The biconcave shape of red blood cells

   a. provides an increased surface area for gas diffusion.
   b. moves the cell membrane closer to hemoglobin.
   c. allows the cell to move through capillaries.
   d. prevents clumping of cells.

7. A person suffering from hypoxia may exhibit a skin and mucous membrane color change known as _______________.

8. Red blood cells cannot reproduce because they lack a _______________.

9. Red blood cell counts are important clinically because they provide information about

   a. blood viscosity.
   b. bone marrow volume.
   c. oxygen carrying capacity.
   d. dietary intake.

10. Damaged red blood cells are destroyed by reticuloendothelial cells called

    a. leukocytes.
    b. macrophages.
    c. neutrophils.
    d. granulocytes.

11. The heme portion of damaged red blood cells is decomposed into iron and

    a. biliverdin.
    b. bilirubin.
    c. bile.

12. In an adult, red blood cells are produced in

    a. the spleen.
    b. red marrow.
    c. yellow marrow.
    d. the liver.

13. Which of the following represents the correct order of appearance of cells in red blood cell production?

    a. erythrocytes, hemocytoblasts, erythroblasts
    b. erythroblasts, hemocytoblasts, erythrocytes
    c. hemocytoblasts, erythrocytes, erythroblasts
    d. hemocytoblasts, erythroblasts, erythrocytes

14. Red blood cell production is stimulated by a hormone, _______________, that is released from the kidney in response to low oxygen concentration.

15. Does statement *a* explain statement *b?* _______________

    a. Vitamin B_{12} and folic acid are necessary to cell growth and reproduction.
    b. The rate of red blood cell reproduction makes this process especially dependent on vitamin B_{12} and folic acid.

16. A lack of vitamin B_{12} is usually due to

    a. dietary deficiency.
    b. a disorder of the stomach lining.
    c. liver damage.
    d. kidney malfunction.

17. An abnormality of hemoglobin metabolism whose symptoms include abdominal pain, fever, headache, and mental changes is called _______________.

18. The most numerous type of white blood cell is the

    a. neutrophil.
    b. eosinophil.
    c. monocyte.
    d. lymphocyte.

19. The white blood cell that has the longest life span is the

    a. basophil.
    b. lymphocyte.
    c. thrombocyte.
    d. eosinophil.

20. The normal white blood cell count is _______________ to _______________ cells per cubic millimeter (mm^3) of blood.

21. The most mobile and active phagocytic lymphocytes are

    a. eosinophils.
    b. neutrophils.
    c. monocytes.
    d. basophils.

22. Does statement *a* explain statement *b?* _______________

    a. All formed elements of blood derive from a common stem cell.
    b. In leukemia, overproduction of white blood cells leads to decreased production of red blood cells and platelets.

23. The blood element concerned with the control of bleeding and the formation of clots is the _______________.

24. Match the functions and characteristics in the first column with the appropriate plasma proteins from the second column.

____ 1. largest molecular size

____ 2. significant in maintaining osmotic pressure

____ 3. transports lipids and fat-soluble vitamins

____ 4. antibody(ies) of immunity

____ 5. plays a part in blood clotting

a. albumins

b. globulins

c. fibrinogen

25. All of the following nutrients are present in plasma except

a. polysaccharides.

b. amino acids.

c. chylomicrons.

d. cholesterol.

26. The liproproteins associated with atherosclerosis are _______________ _______________.

27. The gases that are normally dissolved in plasma include

a. sulfur dioxide.

b. carbon dioxide.

c. oxygen.

d. nitrogen.

28. An increase in the blood level of nonprotein nitrogen can indicate

a. positive nitrogen balance.

b. a kidney disorder.

c. pathologic cell metabolism.

d. overeating.

29. The most abundant plasma electrolytes are

a. calcium.

b. sodium.

c. potassium.

d. chlorides.

30. The vasospasm that occurs in severed blood vessels is due to all of the following except

a. direct stimulation of the vessel wall.

b. release of norepinephrine.

c. stimulation of pain receptors in injured tissue around the vessel.

d. release of serotonin from platelets.

31. A platelet plug begins to form when platelets are

a. exposed to air.

b. exposed to a rough surface.

c. exposed to calcium.

d. crushed.

32. The basic event in the formation of a blood clot is the transformation of a soluble plasma protein, _______________, to a relatively insoluble protein, _______________.

33. Substances believed necessary to activate prothrombin are thought to include

a. calcium ions.

b. potassium ions.

c. phospholipids.

d. glucose.

34. Prothrombin is a plasma protein that is produced by

a. the kidney.

b. the small intestine.

c. the pancreas.

d. the liver.

35. Once a blood clot begins to form, it promotes still more clotting. This is an example of a _______________ feedback system.

36. Widespread activation of the clotting mechanism, which uses up the supply of clotting factors and platelets, is called _______________ _______________ _______________.

37. Laboratory tests used to evaluate the blood coagulation mechanisms are the _______________, _______________, and the _______________ _______________ _______________.

38. Retraction of the clot, pulling the edges of the severed vessel closer together, is due to the action of

a. serum formation.

b. platelets.

c. vitamin K.

d. vitamin C.

39. An enzyme that may be used to dissolve blood clots is _______________.

40. Factors that prevent coagulation in a normal vascular system include all of the following except

a. smooth, unbroken endothelium in blood vessels.
b. a blood plasma protein called antithrombin.
c. heparin.
d. vitamin K.

41. The application of medicinal leeches has been used as an adjunctive therapy to microsurgery to maintain the patency of small veins.

a. True.
b. False

42. The hereditary disease that is almost exclusively male and is due to the lack of one of several clotting factors is _______________.

43. The clumping together of red blood cells when unlike types of blood are mixed is due to antibodies in the plasma and antigens in the

a. thrombocytes.
b. erythrocytes.
c. basophils.
d. eosinophils.

44. A person with type A blood has

a. antigen A and antibody B.
b. antigens A and B.
c. antibodies A and B.
d. neither antibody A nor antibody B.

45. Antibodies for Rh appear

a. spontaneously as an inherited trait.
b. only rarely for poorly understood reasons.
c. only in response to stimulations by Rh antigens.

46. An Rh-negative mother, carrying a fetus who is Rh-positive, may have an infant with a blood problem called _______________ _______________.

47. An Rh-negative mother who delivers an Rh-positive baby is given _______________ within 72 hours of delivery to prevent the condition in question 46.

48. Which of the following blood types is a universal donor?

a. O
b. AB
c. A
d. B

STUDY ACTIVITIES

I. Definition of Key Terms

Define the following key terms used in this chapter.

albumin

antibody

antigen

basophil

coagulation

embolus

eosinophil

erythrocyte

erythropoietin

fibrinogen

globulin

hemostasis

leukocyte

lymphocyte

macrophage

monocyte

neutrophil

plasma

platelet

thrombus

II. Introduction (p. 521)

A. What is the advantage of using stem cells rather than bone marrow transplantation to treat individuals who have diseases that affect blood cells?

B. Why is blood from umbilical cords used for this purpose?

III. Blood and Blood Cells (pp. 521–32)

A. Answer the following concerning the volume and composition of blood.

1. List the solids of the blood.

2. What is the blood volume of an average (70-kilogram) male?

3. What is the hematocrit?

4. What part of blood tissue is plasma?

5. Describe the origin of blood cells.

B. Answer the following questions concerning red blood cells.

1. The shape of a red blood cell is a _______________ _______________.
2. How does their shape enhance the function of red blood cells?

3. How does the lack of a nucleus affect red blood cell function?

4. Describe sickle cell anemia.

5. Why is hydroxyurea effective in the treatment of sickle cell anemia?

C. Answer these questions concerning red blood cell counts.

1. What are the normal red cell counts for a man, a woman, and a child?

2. How are the plasma volume and hematocrit related?

D. Answer the following concerning the destruction of red blood cells.

1. How are red blood cells damaged?

2. Damaged red blood cells are destroyed by cells called _______________, located in the _______________ and _______________.
3. Describe how the body recycles hemoglobin.

4. What is hemochromatosis?

E. Answer the following concerning red blood cell production and its control.

1. Where are red blood cells produced?

2. Describe the process of red blood cell production.

3. How is the production of red blood cells controlled?

4. What is the effect of increasing the sensitivity of reticulocytes to erythropoietin?

F. Answer the following concerning dietary factors affecting red blood cell production.

1. What is the role of folic acid and vitamin B_{12} in red blood cell production?

2. What is the role of intrinsic factor in red blood cell production?

3. How does pregnancy affect red blood cell production?

G. Answer the following concerning red blood cell disorders.

1. Fill in the following chart.

Types of anemia

Type	**Character of anemia**	**Cause**
Pernicious		
Aplastic		
Hemolytic		
Iron deficiency		
Sickle cell disease		
Thalassemia		

2. Describe the pathologic events and symptoms of porphyria.

H. Fill in the following chart.

Types of white blood cells

White blood cell	Description	Percent of total	Function
Granulocytes			
Neutrophil			
Eosinophil			
Basophil			
Agranulocytes			
Monocyte			
Lymphocyte			

I. Answer the following concerning the functions of white blood cells.

1. How do white cells protect the body against microorganisms?

2. What is diapedesis?

3. Review and describe the inflammatory reaction. Include the role of histamine and the phenomenon of chemotaxis.

J. Answer these questions concerning white blood cell counts.

1. What is a normal white cell count?

2. What causes an increase or a decrease in white blood cells?

3. What is a differential white blood cell count?

4. What white cell count is decreased in AIDS patients?

K. Answer the following questions about leukemia.

1. How does leukemia affect white blood cell function?

2. Compare myeloid and lymphatic leukemias.

3. What types of leukemia are curable?

L. Describe the structure and function of platelets.

IV. Blood Plasma (pp. 532–37)

A. Fill in the following chart.

Plasma proteins

Protein	Description	Percentage of total	Function
Albumins			
Globulins			
Fibrinogen			

B. How does the concentration of plasma proteins affect water balance?

C. Answer the following concerning nutrients and gases.

1. What nutrients are found in plasma?

2. Describe the lipids found in plasma. Be sure to include phospholipids and chylomicrons.

3. What gases are found in plasma? How are their levels evaluated?

4. Describe the nonprotein nitrogenous substances found in plasma.

5. What factors can provoke an increase in these substances?

D. Answer these questions concerning plasma electrolytes.

1. What electrolytes are found in plasma?

2. What is the function of electrolytes?

V. **Hemostasis** (pp. 537–42)

A. Answer the following concerning hemostasis and blood vessel spasm.

1. List the mechanisms of hemostasis.

2. Are these mechanisms most likely to be successful in blood loss from small or from large vessels?

3. What are the stimuli for vasospasm?

B. How is a platelet plug formed?

C. Describe the major events in the blood clotting mechanism.

D. Compare the extrinsic and intrinsic clotting mechanisms.

E. What happens to a blood clot?

F. Describe the mechanisms that prevent coagulation.

G. What drugs can be used to dissolve clots in the body?

H. Answer these questions concerning coagulation disorders.

1. What kind of abnormal bleeding is associated with thrombocytopenia?

2. What is hemophilia? Compare it to von Willebrand disease.

3. What is a thrombus? What conditions predispose a person to the formation of thrombi?

4. What is an embolus?

5. Why are leeches used in modern medical practice?

VI. Blood Groups and Transfusions (pp. 542–47)

A. Answer the following concerning antigens and antibodies.

1. What are antigens and antibodies?

2. Antigens are present in the _______________ _______________, antibodies in the _______________.

B. Answer the following concerning the ABO blood group.

1. Describe the basis for ABO blood types.

2. Why is it unsafe to mix different blood types?

C. Answer the following concerning the Rh blood group.

1. What is the Rh factor?

2. How does the Rh factor differ from A, B, and O antigens?

3. Describe how erythroblastosis fetalis develops both in utero and after birth.

4. How can this problem be prevented?

5. Describe a transfusion reaction (agglutination reaction).

D. How does crossmatching prevent transfusion reactions?

VII. Clinical Focus Question

Your uncle Bob is planning to have an elective hip replacement and his surgeon has discussed the possibility of Uncle Bob storing several units of his blood in case he needs a transfusion during surgery. Your Uncle Bob asks you what you think about this. How would you respond?

When you have completed the study activities to your satisfaction, retake the mastery test and compare your performance with your initial attempt. If there are still areas you do not understand, repeat the appropriate study activities.

CHAPTER 15
CARDIOVASCULAR SYSTEM

OVERVIEW

This chapter deals with the system that transports blood to and from cells—the cardiovascular system. It identifies the major organs of the cardiovascular system and explains the function of each organ (objective 1). In this chapter the location, structure, and function of the parts of the heart and the major types of blood vessels are discussed (objectives 2 and 5). It explains the pulmonary, systemic, and coronary circuits of the cardiovascular system and the major vessels in each circuit (objectives 3, 8 and 9). This chapter discusses the cardiac cycle and its control (objective 4). In addition, it describes how blood pressure is created and controlled, and the mechanism that aids in the return of venous blood to the heart (objectives 6 and 7).

Study of the cardiovascular system is essential for understanding how each part of the body is supplied with the materials it needs to sustain life.

CHAPTER OBJECTIVES

After you have studied this chapter, you should be able to:

1. Name the organs of the cardiovascular system and discuss their functions.
2. Name and describe the locations of the major parts of the heart and discuss the function of each part.
3. Trace the pathway of the blood through the heart and the vessels of the coronary circulation.
4. Discuss the cardiac cycle and explain how it is controlled.
5. Compare the structures and functions of the major types of blood vessels.
6. Describe the mechanisms that aid in returning venous blood to the heart.
7. Explain how blood pressure is produced and controlled.
8. Compare the pulmonary and systemic circuits of the cardiovascular system.
9. Identify and locate the major arteries and veins of the pulmonary and systemic circuits.

FOCUS QUESTION

As you run up the stairs from your laundry to your kitchen to answer the phone, your heart beats a little faster. How do your cardiovascular and respiratory systems work together to supply cells with oxygen at all levels of activity?

MASTERY TEST

Now take the mastery test. Do not guess. Some questions may have more than one correct answer. As soon as you complete the test, correct it. Note your successes and failures so that you can read the chapter to meet your learning needs.

1. Rheumatic fever damages the
   a. walls of the ventricles.
   b. walls of the atria.
   c. membranes that cover the heart.
   d. valves of the heart.
2. The heart is a cone-shaped, muscular pump located within the _______________.
3. The apex of the heart is located
   a. beneath the sternum, in the fifth intercostal space.
   b. in the second intercostal space, below the sternum.
   c. in the fourth intercostal space, in the midaxillary line.
   d. in the fifth intercostal space, about 3 inches left of the midline.

4. The visceral pericardium is also known as the
   a. epicardium.
   b. myocardium.
   c. endocardium.

5. The most life-threatening aspect of pericarditis is
   a. pain.
   b. the possibility of generalized sepsis.
   c. damage to the muscle of the heart.
   d. restriction of heart movements.

6. Purkinje fibers are located in the
   a. epicardium.
   b. myocardium.
   c. endocardium.
   d. parietal pericardium.

7. The upper chambers of the heart are the right and left _______________; the lower chambers are the right and left _______________.

8. The hormone atrial natriuretic peptide (ANP) is released by the muscles of the atria in response to excessive _______________ of the atria.

9. The vessels that empty into the upper right chamber of the heart are the
   a. inferior and superior venae cavae.
   b. pulmonary veins.
   c. pulmonary arteries.
   d. coronary sinuses.

10. The valve between the chambers of the right side of the heart is the _______________ valve.
   a. semilunar
   b. bicuspid (mitral valve)
   c. tricuspid
   d. aortic

11. The structures that prevent the mitral and tricuspid valves from swinging into the atria during ventricular contraction are the
   a. cordae tendineae.
   b. papillary muscles.
   c. endocardium.
   d. Purkinje fibers.

12. Which of the following statements about the muscle walls of the right and left heart chambers is/are true?
   a. The muscle walls of the right and left heart are the same size.
   b. The muscle walls of the atria are thicker than those of the ventricles.
   c. The muscle walls of the right heart are thinner than those of the left heart.
   d. The muscle walls of the left ventricle are 2-3 times thicker than those of the right ventricle.

13. The valve that separates the left ventricle from the aorta is the ____________ valve.

14. A clicklike heart sound over the apex at the end of ventricular contraction indicates
   a. mitral valve prolapse.
   b. aortic stenosis.
   c. endocarditis.
   d. a normal heart.

15. Dense fibrous rings that surround the valves of the heart and the superior portion of the interventricular septum make up the _______________ _______________.

16. Blood is supplied to the heart by the right and left _______________ _______________.

17. The branch of the left coronary artery that supplies the walls of both ventricles with blood is the _______________ artery.
   a. circumflex
   b. anterior interventricular
   c. posterior interventricular
   d. marginal

18. The condition that occurs when cardiac muscle dies due to lack of circulation is called _______________ _______________.

19. The hearts of large numbers of African American adults fail due to deposition of a protein called _______________ in the tissues of the heart.

20. Blood flow to cardiac muscle (increases/decreases) during ventricular contraction.

21. Atrial contraction, while the ventricles relax, followed by ventricular contraction, while the atria relax, is known as the _______________ _______________.

22. Heart sounds heard with a stethoscope are produced by
   a. contraction of the muscle of the heart.
   b. the flow of blood through the heart.
   c. the opening and closing of the heart valves.
   d. changes in the velocity of blood flow through the heart.
23. A mass of merging cells that function as a unit is called
   a. smooth muscle.
   b. a functional syncytium.
   c. the sinoatrial node.
   d. the cardiac conduction system.
24. The cells that initiate the stimulus for contraction of the heart muscle are located in the
   a. sinoatrial node.
   b. atrioventricular node.
   c. Purkinje fibers.
   d. bundle of His.
25. The conduction system cells that are continuous with cardiac muscle cells are
   a. sinoatrial node cells.
   b. atrioventricular node cells.
   c. Purkinje fibers.
   d. cells of the bundle of His.
26. The effect of vagus nerve stimulation on the heart is to (decrease/increase) the heart rate.
27. Abnormalities in the concentration of which of the following ions are likely to interfere with the contraction of the heart?
   a. chloride
   b. potassium
   c. calcium
   d. sodium
28. In the ECG, atrial contraction is represented by the
   a. P wave.
   b. QRS complex.
   c. T wave.
   d. U wave.
29. When a heart rate is 150 beats per minute, the rhythm is said to be
   a. bradycardia.
   b. flutter.
   c. fibrillation.
   d. tachycardia.
30. If an individual has a heart rate of 50 beats per minute, which of the following is probably the pacemaker?
   a. S–A node
   b. A–V node
   c. Purkinje fibers
31. An individual who has a damaged cardiac conduction system may be treated with
   a. defibrillation.
   b. an artificial pacemaker.
   c. a coronary artery bypass procedure.
   d. exercise and diet.
32. The vessel that participates directly in the exchange of substances between the cell and the blood is the
   a. arteriole.
   b. artery.
   c. capillary.
   d. venule.
33. The arterial wall layer that is made up of epithelial tissue and connective tissue rich in elastic and collagenous tissue is the
   a. tunica intima.
   b. tunica media.
   c. tunica adventitia.
34. Substances that inhibit platelet aggregation and cause vasodilation are secreted by the _______________ of blood vessels.
35. The amount of blood that flows into capillaries is regulated by
   a. constriction and dilation of capillaries.
   b. arterioles.
   c. the amount of intercellular tissue.
   d. precapillary sphincters.
36. The blood/brain barrier is created by the relative _______________ of capillaries in the brain.
37. The transport mechanisms used by the capillaries are _______________, _______________, and _______________.
38. Blood pressure is highest in a(n)
   a. artery.
   b. arteriole.
   c. capillary.
   d. vein.

39. The most important transport mechanism in the capillary bed is

a. osmosis.
b. diffusion.
c. filtration.
d. active transport.

40. Plasma proteins help retain water in the blood by maintaining

a. osmotic pressure.
b. hydrostatic pressure.
c. a vacuum.

41. Molecules of oxygen, carbon dioxide, and glucose move across the capillary wall by ______________.

42. Swelling occurs with tissue injury due to

a. breakdown of capillary walls.
b. constriction of precapillary sphincters.
c. increased permeability of capillary walls.
d. constriction of venules.

43. The middle layer of the walls of veins differs from that of arteries in that it

a. contains more connective tissue.
b. contains less smooth muscle.
c. is thicker.
d. contains some striated muscle.

44. Blood in veins is kept flowing in one direction by the presence of ______________.

45. An inflammation of the vein associated with the formation of a clot is called

a. phlebitis.
b. thrombophlebitis.
c. thrombitis.
d. phlebosis.

46. Increased venous pressure with dilation of veins is characteristic of ______________ ______________.

47. The maximum pressure in the artery, occurring during ventricular contraction, is

a. diastolic pressure.
b. systolic pressure.
c. mean arterial pressure.
d. pulse pressure.

48. The difference between the systolic and diastolic pressure is the ______________ pressure.

49. The amount of blood pushed out of the ventricle with each contraction is called ______________ ______________.

50. List the five factors that influence blood pressure.

51. Cardiac output is calculated by multiplying ______________ ______________ by ______________ ______________.

52. Starling's law is related to which of the following cardiac structures?

a. interventricular septum
b. conduction system
c. muscle fibers
d. heart valves

53. When the vasoreceptors in the aorta and carotid artery sense an increase in blood pressure, the medulla will relay (sympathetic/parasympathetic) impulses.

54. Peripheral resistance is maintained by increasing or decreasing the size of

a. capillaries.
b. arterioles.
c. venules.

55. When the underlying cause of high blood pressure is known, the individual is said to have ______________ hypertension.

56. Angiotensin increases blood pressure by

a. promoting vasoconstriction.
b. increasing blood volume.
c. increasing cardiac output.
d. increasing viscosity.

57. Hypertension may set up a positive feedback situation by

a. increasing the workload of the heart.
b. increasing peripheral resistance.
c. enhancing the development of atherosclerosis.
d. decreasing blood flow to the kidney.

58. Venous blood flow is maintained solely by muscle contraction so that blood is "milked" forward as veins are compressed.

a. True

b. False

59. The central venous pressure is the pressure in the

a. left atrium.

b. right atrium.

c. right ventricle.

d. left ventricle.

60. Which of the following vessels carries deoxygenated blood?

a. aorta

b. innominate artery

c. basilar artery

d. pulmonary artery

61. The pulmonary veins enter the ______________ ______________.

62. The aortic bodies containing pressoreceptors and chemoreceptors are located in the

a. ascending aorta.

b. aortic arch.

c. aortic sinuses.

d. descending aorta.

63. The ring of arteries at the base of the brain is called the ______________ ______________ ______________.

64. The veins that drain the abdominal viscera empty into a unique venous system called the ______________ ______________ ______________.

65. Venous blood flow is maintained by all but which of the following factors?

a. blood pressure

b. skeletal muscle contraction

c. vasoconstriction of veins

d. respiratory movements

STUDY ACTIVITIES

I. Definition of Key Terms

Define the following key terms used in this chapter.

arteriole

atrium

cardiac conduction system

cardiac cycle

cardiac output

diastole

electrocardiogram

endocardium

epicardium

functional syncytium

myocardium

pacemaker

pericardium

peripheral resistance

pulmonary circuit

sphygmomanometer

systemic circuit

systole

vasoconstriction

vasodilation

ventricle

venule

viscosity

II. Introduction and Structure of the Heart (pp. 554–65)

A. Answer these questions about the effects of rheumatic heart disease.

1. What cardiac structures are affected by rheumatic heart disease?

2. What are the symptoms of rheumatic heart disease?

3. How is rheumatic heart disease treated?

B. Answer the following concerning the cardiovascular system and the location of the heart.

1. What is the function of the cardiovascular system?

2. Describe the location of the heart precisely.

3. Locate the apical heartbeat.

C. Answer the following concerning the coverings of the heart.

1. The heart is enclosed by a double-layered _______________.
2. What is the function of the fluid in the pericardial space?

3. What are the results of an inflammation of the pericardium?

D. Answer the following concerning the wall of the heart.

1. Label the accompanying drawing: myocardium, endocardium, epicardium.

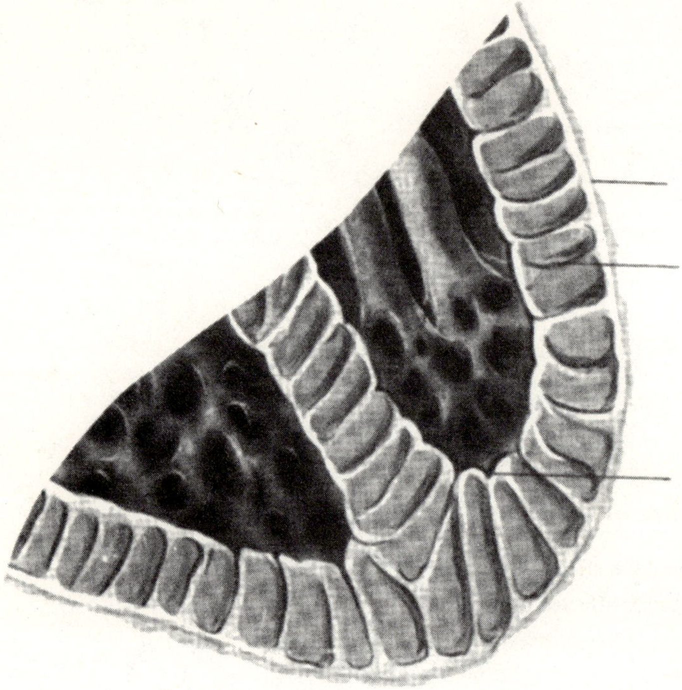

2. Describe the structure and function of each of the labeled portions of the wall of the heart.

E. Answer the following concerning heart chambers and valves.

1. List the chambers of the heart.

2. What is the function of atrial natriuretic peptide?

3. How does it achieve its effect on circulation?

4. Why is the muscle wall of the right side of the heart smaller than that of the left side of the heart?

5. Describe the cardiac events, symptoms, and treatment of mitral valve prolapse.

6. Label the structures in the accompanying drawing: interventricular septum, right and left ventricles, right and left atria, superior and inferior venae cavae, aorta, tricuspid valve, pulmonary valve, aortic valve, pulmonary trunk, bicuspid valve, right and left pulmonary veins, right and left pulmonary arteries, chordae tendineae, papillary muscle, opening of coronary sinus.

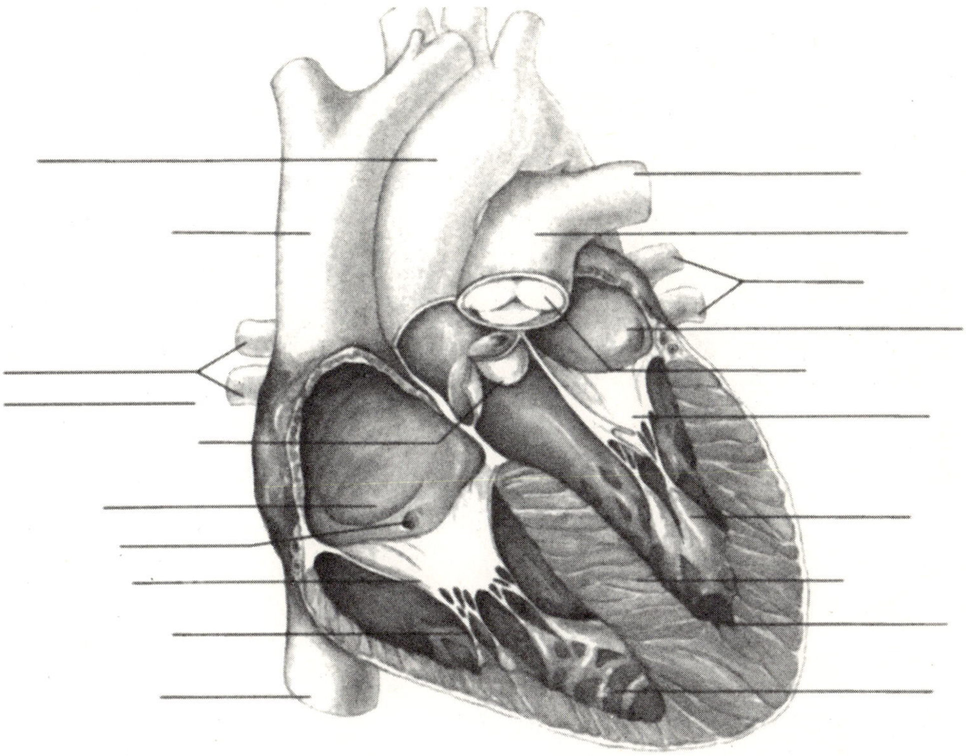

7. What vessels bring blood to the right atrium?

8. What vessels bring blood to the left atrium?

9. What is the function of the valves of the heart?

10. When the cusps of the mitral valve stretch and bulge upward into the atrium during ventricular contraction, the individual is said to have _______________ _______________ _______________.

F. Describe the skeleton of the heart.

G. Trace the path of the blood through the heart. Include all valves.

H. Answer the following concerning the blood supply to the heart.

1. Describe the blood supply to the heart.

2. Is the blood flow to the myocardium greatest during systole or diastole? What is the clinical significance of this fact?

3. What is the result when the heart is deprived of a blood supply?

4. Describe familial amyloidosis including the population at risk and method of diagnosis.

III. Heart Actions (pp. 565–73)

A. Answer these questions concerning the cardiac cycle.

1. What events make up a cardiac cycle?

2. What produces the heart sounds heard with a stethoscope?

3. What structures of the heart can be assessed by heart sounds?

B. Describe the characteristics of cardiac muscle fibers.

C. Answer the following concerning the cardiac conduction system.

1. Label the parts of the cardiac conduction system in the accompanying drawing: S–A node, A–V node, A–V bundle, Purkinje fibers, interventricular septum, interatrial septum, left bundle branch.

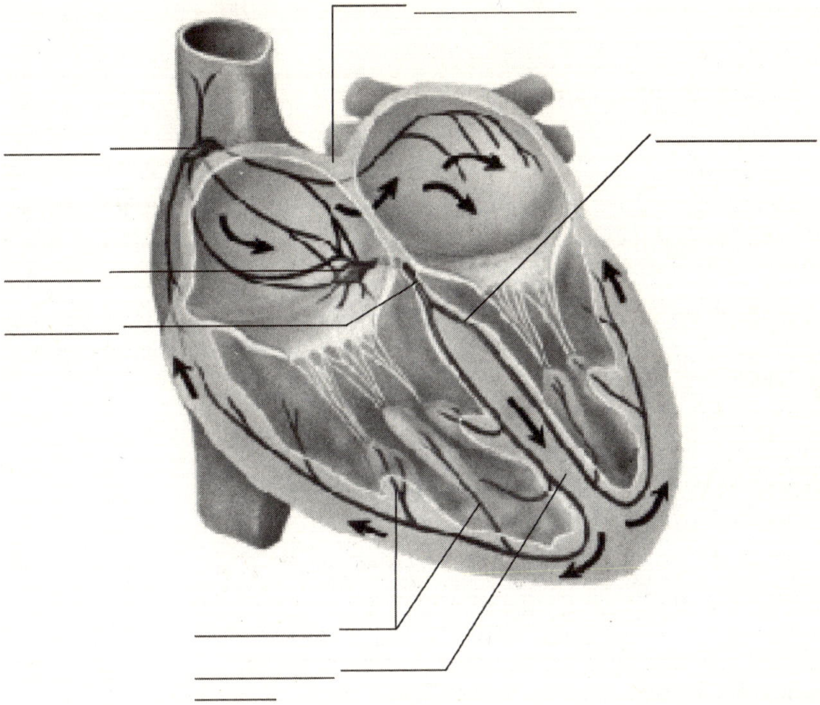

2. Describe the conduction of an impulse from the S-A node to contraction of the myocardium.

3. In what part of the ventricle does contraction begin? What effect does this have on the movement of blood through the ventricle?

D. Answer the following concerning the electrocardiogram.

1. A recording of electrical changes in the myocardium is a(n) ______________.

2. How is this recording obtained?

3. What events in the cardiac cycle are represented by each of the following: P wave, QRS complex, T wave?

4. What is the significance of a lengthened P-R interval?

E. Answer these questions concerning regulation of the cardiac cycle.

1. How is the heart regulated by autonomic reflexes?

2. How is the heart affected by the following: potassium, calcium?

3. How does temperature affect the heart?

F. Answer the following concerning heart actions.

1. Match the arrhythmias in the first column with their correct descriptions in the second column.

____ 1. tachycardia	a.	rapid, uncoordinated heartbeat
____ 2. bradycardia	b.	heartbeat over 100 beats per minute
____ 3. premature heartbeat	c.	heartbeat over 250 beats per minute
____ 4. flutter	d.	heartbeat under 60 beats per minute
____ 5. fibrillation	e.	a beat that occurs earlier than expected in the cardiac cycle

2. What is a conduction disorder?

3. Compare a pacemaker and a pacemaker-cardioverter-defibrillator.

IV. Blood Vessels (pp. 573–93)

A. Answer the following concerning the cardiovascular system and arteries.

1. Describe the closed circuit formed by blood vessels.

2. What is the structure of the wall of arteries?

3. How does the endothelium prevent blood clotting and contribute to the control of blood pressure?

4. How is the structure of arterioles different from that of arteries?

B. Answer the following concerning capillaries.

1. What is the structure of capillaries?

2. What determines the density of capillaries within tissue?

3. Explain how the variation in the permeability of capillaries occurs.

4. How is the distribution of blood in the various capillary pathways regulated?

5. Describe the following transport mechanisms in the capillary: diffusion, filtration, osmosis.

6. Describe the events of right-sided heart failure.

C. Answer these questions concerning veins.

1. What is the structure of a vein?

2. In what ways is this different from the structure of an artery?

3. How do veins function as blood reservoirs?

D. Match the blood vessel disorders in the first column with the correct descriptions in the second column.

____ 1. atherosclerosis	a. inflammation of a vein
____ 2. arteriosclerosis	b. abnormal dilation of a superficial vein
____ 3. varicose veins	c. characterized by an accumulation of fatty plaques
____ 4. phlebitis	d. degenerative changes that lead to a loss of elasticity

V. **Blood Pressure** (pp. 582–92)

A. Answer these questions concerning arterial blood pressure.

1. What is blood pressure?

2. What cardiac events are related to systolic and diastolic arterial pressure?

3. What is a pulse?

B. Answer these questions concerning the measurement of arterial blood pressure.

1. How is blood pressure measured?

2. What are pulse pressure and mean arterial pressure?

C. How does each of these factors influence arterial blood pressure: heart action, blood volume, peripheral resistance, viscosity?

D. Answer the following concerning the control of blood pressure.

1. How is cardiac output calculated?

2. Discuss the mechanical, neural, and chemical factors that affect cardiac output.

3. How is peripheral resistance regulated?

E. Answer the following concerning high blood pressure.

1. Why does arteriosclerosis lead to high blood pressure?

2. What mechanism in kidney disease leads to high blood pressure?

3. Discuss the results of high blood pressure.

F. What is central venous pressure?

G. How do factors such as skeletal muscle contraction, breathing movements, and vasoconstriction of veins influence venous blood flow?

H. How does the cardiovascular system respond to exercise?

VI. **Paths of Circulation** (pp. 592–95)

A. Trace a drop of blood through the pulmonary circuit, noting whether it is oxygenated or deoxygenated.

B. What is the function of the systemic circuit?

VII. **Arterial System and Venous System** (pp. 595–614)

A. List the principal branches of the aorta in the order in which they arise, beginning with those that arise from the aortic arch.

B. Describe the circulation to the head, neck, and brain.

C. List the arteries that supply the shoulder and upper limb.

D. List the arteries that supply the thoracic and abdominal walls.

E. List the arteries that supply the pelvis and lower limbs.

F. Describe the relationship between the arterial and venous vessels.

G. Why is control of cholesterol essential to cardiovascular health?

VIII. Clinical Focus Question

David is a fifty-two-year-old overweight accountant who has had chest pain occasionally for several months, usually related to increased exercise. His cardiologist has told him he has coronary artery disease.

A. What are the risk factors for heart disease?

B. What foods should David avoid?

C. How might biofeedback be helpful to David?

When you have completed the study activities to your satisfaction, retake the mastery test and compare your performance with your initial attempt. If there are still areas you do not understand, repeat the appropriate study activities.

CHAPTER 16
LYMPHATIC SYSTEM AND IMMUNITY

OVERVIEW

The lymphatic system has two major functions: (1) it helps maintain fluid balance in the tissues of the body, and (2) it has a major role in the defense against infection. This chapter describes the general functions of the lymphatic system (objective 1). In the discussion of fluid balance, this chapter describes the major lymphatic pathways, lymph formation, and circulation (objectives 2 4). In discussing the defense function, it explains lymph nodes, their location, and their function (objectives 5 and 6). Specific and nonspecific body defenses, and the functions of the thymus, spleen, lymphocytes, and immunoglobulins (objectives 7–10) are discussed. Various types of immune responses, including primary and secondary responses, active and passive responses, allergic reactions, and tissue rejection reactions (objectives 11–13), are also explained.

The study of the lymphatic system completes your knowledge of how fluid is transported to and from tissues. Knowledge of the immune mechanisms of the lymphatic system is the basis for understanding how the body defends itself against specific kinds of threat.

CHAPTER OBJECTIVES

After you have studied this chapter, you should be able to:

1. Describe the general functions of the lymphatic system.
2. Describe the location of the major lymphatic pathways.
3. Describe how tissue fluid and lymph form and explain the function of lymph.
4. Explain how lymphatic circulation is maintained and describe the consequence of lymphatic obstruction.
5. Describe a lymph node and its major functions.
6. Describe the location of the major chains of lymph nodes.
7. Discuss the functions of the thymus and spleen.
8. Distinguish between specific and nonspecific defenses, and provide examples of each.
9. Explain how two major types of lymphocytes are formed, activated, and how they function in immune mechanisms.
10. Name the major types of immunoglobulins and discuss their origins and actions.
11. Distinguish between primary and secondary immune responses.
12. Distinguish between active and passive immunity.
13. Explain how allergic reactions, tissue rejection reactions, and autoimmunity arise from immune mechanisms.

FOCUS QUESTION

Several days after falling and scraping your knee, you notice that the area around your knee is swollen and that you have several enlarged lymph nodes in your groin. How are these signs related to the function of the lymphatic system in maintaining fluid balance and responding to injury and infection?

MASTERY TEST

Now take the mastery test. Do not guess. Some questions may have more than one correct answer. As soon as you complete the test, correct it. Note your successes and failures so that you can read the chapter to meet your learning needs.

1. Organ transplants from living, related donors are more successful than transplants from nonrelatives.
   a. True
   b. False

2. The smallest vessels in the lymphatic system are called _______________ _______________. The largest vessels are called _______________ _______________.

3. The lymphatic vessels in the villi of the small intestine, called _______________, are involved in the absorption of _______________.

4. The largest lymph vessel is the
   a. lumbar trunk.
   b. thoracic duct.
   c. lymphatic duct.
   d. intestinal trunk.

5. Lymph rejoins the blood and becomes part of the plasma in
   a. lymph nodes.
   b. the right and left subclavian veins.
   c. the inferior and superior venae cavae.
   d. the right atrium.

6. Tissue fluid originates from
   a. the cytoplasm of cells.
   b. lymph fluid.
   c. blood plasma.

7. Lymph formation is most directly dependent on
   a. increasing osmotic pressure in tissue fluid.
   b. a blood pressure of at least 100/60.
   c. a sufficient volume of tissue fluid to create a pressure gradient between tissue and lymph capillaries.
   d. diminished peripheral resistance in veins.

8. The function(s) of lymph are to
   a. recapture protein molecules lost in the capillary bed.
   b. form tissue fluid.
   c. transport foreign particles to lymph nodes.
   d. recapture electrolytes.

9. The mechanisms that move lymph through lymph vessels are similar to those that move blood through (arteries/veins).

10. The flow of lymph is greatest during periods of
   a. physical exercise.
   b. isometric exercise of skeletal muscle.
   c. dream sleep.
   d. REM sleep.

11. Obstruction of lymph circulation will lead to _______________.

12. Lymph nodes are shaped like
   a. almonds.
   b. peas.
   c. kidney beans.
   d. convex disks.

13. Compartments within the node contain dense masses of
   a. epithelial tissue.
   b. cilia.
   c. oocytes.
   d. lymphocytes.

14. Inflammation of a lymph node is known as _______________.

15. Clumps of lymph nodes in mucosa of the ileum are called _______________ _______________.

16. An infection in the toe would result in enlarged lymph nodes in the
   a. axilla.
   b. inguinal region.
   c. pelvic cavity.
   d. abdominal cavity.

17. Which of the following types of cells are produced by lymph nodes?
   a. leukocytes
   b. lymphocytes
   c. eosinophils
   d. basophils

18. The thymus is located in the
   a. posterior neck.
   b. mediastinum.
   c. upper abdomen.
   d. left pelvis.

19. Which of the following statements about the thymus is/are true?

a. The thymus tends to increase in size with age, as glandular tissue is replaced by fat and connective tissue.

b. The thymus is relatively large during infancy and childhood.

c. The thymus produces a substance called thymosin that seems to stimulate the development of lymphatic tissue.

d. The thymus is a firm, multilobed structure.

20. The largest of the lymphatic organs is the _______________.

21. Which of the following statements about the spleen is/are true?

a. The spleen is located in the lower left quadrant of the abdomen.

b. The spleen functions in the body s defense against infection.

c. The structure of the spleen is exactly like that of a lymph node.

d. Splenic pulp contains large phagocytes and macrophages on the lining of its venous sinuses.

22. Agents that enter the body and cause disease are called _______________.

23. A disease characterized by the alteration of species resistance to microorganisms is

a. cancer.

b. AIDS.

c. malaria.

d. viral pneumonia.

24. The skin is an example of which of the following defense mechanisms?

a. immunity

b. inflammation

c. mechanical barrier

d. phagocytosis

25. Which of the following characteristics of the stomach enables it to act as a defense mechanism?

a. low pH

b. presence of lysozyme

c. presence of amylase

d. presence of pepsin

26. The enzyme lysozyme, which has antibacterial ability, is present in which of the following body fluids?

a. sweat

b. blood

c. tears

d. urine

27. List the four major symptoms of inflammation.

28. The accumulation of white blood cells, bacterial cells, and damaged tissue cells creates

a. exudate.

b. pus.

c. a scab.

29. The most active phagocytes in the blood are _______________ and _______________.

30. Phagocytes that remain fixed in position within various organs are called

a. neutrophils.

b. monocytes.

c. macrophages.

31. Macrophages are located in the lining of blood vessels in the bone marrow, liver, spleen, and lymph nodes; they form the _______________ system.

32. The resistance to specific foreign agents in which certain cells recognize foreign substances and act to destroy them is _______________.

33. Some undifferentiated lymphocytes migrate to the _______________ _______________ where they undergo changes and are then called T lymphocytes.

34. Foreign proteins to which lymphocytes respond are called _______________.

35. Lymphocytes seem to be able to recognize specific foreign proteins because

a. of changes in the nucleus of the lymphocyte.

b. the cytoplasm of the lymphocyte is altered.

c. there are changes in the permeability of the cell membrane of the lymphocyte.

d. of the presence of receptor molecules on lymphocytes, which fit the molecules of antigens.

36. B lymphocytes respond to foreign protein by

a. phagocytosis.
b. interacting directly with pathogens.
c. producing antigens.
d. producing antibodies.

37. T cells are responsible for _______________ immunity.

38. The antibodies produced by B cells make up the _______________ _______________ fraction of plasma.

39. The immunoglobulin that crosses the placenta from the mother to the fetus is immunoglobulin

a. A.
b. G.
c. M.

40. Which of the following actions of antibodies is most important in protecting the body against infection?

a. agglutination
b. precipitation
c. complement
d. neutralization

41. An accessory cell is necessary to activate (B cells/T cells).

42. In which of the following ways are primary and secondary immune responses different?

a. Primary responses are more important than secondary responses.
b. Primary responses produce more antibodies than secondary responses.
c. A primary response is a direct response to an antigen; a secondary response is indirect.
d. A primary response is the initial response to an antigen; a secondary response is all subsequent responses to that antigen.

43. A person who receives ready-made antibodies develops artificially acquired _______________ immunity.

44. Age-related decline in the competence of the immune system is due to the loss of _______________.

45. The retrovirus that causes Acquired Immune Deficiency Syndrome is transmitted by

a. the airborne route.
b. contact with infected articles.
c. inoculation with infected blood.
d. unknown means.

46. An individual with special abilities to carry on an abnormal immune reaction to a usually harmless substance is said to have

a. an increased potential for cancer.
b. an allergic reaction.
c. a collagen disease.

47. A common problem following organ transplant is _______________ _______________.

48. A major problem with the use of immunosuppressive drugs is _______________.

49. When tolerance to self-substance is lost, and the immune response is directed against the individual s own tissue, the individual is said to have an _______________ disease.

STUDY ACTIVITIES

I. Definition of Key Terms

Define the following key terms used in this chapter.

allergen

antibody

antigen

clone

complement

hapten

immunity

immunoglobulin

interferon

lymph

lymphatic pathway

lymph node

lymphocyte

macrophage

pathogen

reticuloendothelial tissue

spleen

thymus

vaccine

II. Introduction and Lymphatic Pathways (pp. 621–25)

A. Why are organ transplants from related donors more likely to be successful than those from nonrelated donors?

B. Describe the relationship between the lymphatic system and the cardiovascular system.

C. What are the components of a lymphatic pathway?

D. How are lacteals related to lymph capillaries?

III. Tissue Fluid and Lymph (p. 625)

A. How is tissue fluid formed? Include the composition of tissue fluid and transport mechanisms used.

B. How is lymph formed? How is lymph formation related to tissue-fluid formation?

C. What is the function of lymph?

D. How does the structure of the walls of lymph capillaries prompt the movement of fluid from tissue into lymph capillaries?

IV. Lymph Movement and Lymph Nodes (pp. 626–28)

A. Describe the forces responsible for the circulation of lymph.

B. Why are lymph nodes removed in many surgical procedures to remove a cancer?

C. How can surgery obstruct lymph movement?

V. Lymph Nodes (pp. 626–28)

A. Label the structures in the following drawing: afferent lymphatic vessel, sinus, nodule, efferent lymphatic vessel, hilum, capsule, subcapsule, germinal center, artery, vein, trabecula. In addition, use arrows to indicate the direction of lymph flow through the lymph node.

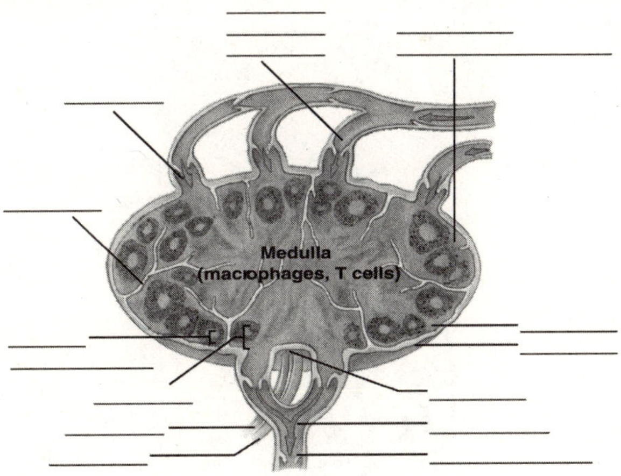

B. Answer the following concerning the locations of lymph nodes.

1. List the locations of lymph nodes.

2. Use yourself or a partner to locate the major chains of lymph nodes. (Will you be able to feel these nodes in an individual who is free from infection?)

3. Inflammation of lymph vessels is ______________. Inflammation of a lymph node is ______________.

C. Why are the supratrochlear lymph nodes often enlarged in children?

D. What is the function of lymph nodes?

VI. Thymus and Spleen (pp. 628–30)

A. Describe the location, structure, and function of the thymus gland.

B. Answer the following concerning the spleen.

1. Where is the spleen located?

2. Label the structures in the following diagram: capsule, venous sinuses, red pulp, white pulp, connective tissue, artery of pulp, spleen, splenic artery, and splenic vein.

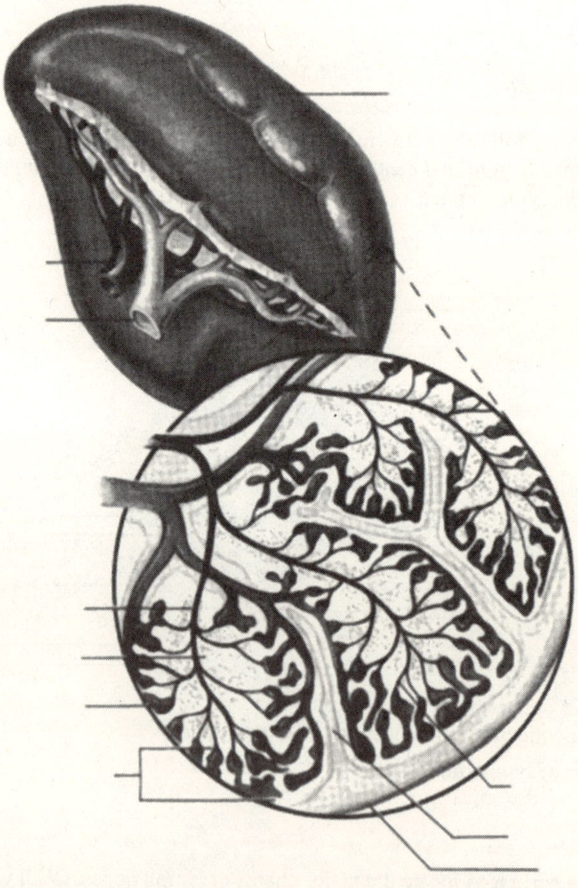

3. What characteristics of the spleen allow it to function in the defense against foreign protein?

4. In what circumstances may the spleen revert to the manner in which it functioned during fetal life?

VII. Body Defenses Against Infection (p. 630)

A. What kinds of agents cause disease?

B. What two major types of defense mechanisms prevent disease?

VIII. Nonspecific Defenses (pp. 630–32)

A. What is species resistance?

B. What structures function as mechanical barriers?

C. What enzymes help resist infection by acting as chemical barriers?

D. What is interferon? How does interferon work?

E. What is the role of fever in helping the body to overcome an infection?

F. Answer the following concerning inflammation.

1. What is inflammation?

2. Explain the reason for the major signs and symptoms of inflammation.

   redness

   swelling

   heat

   pain

3. Describe how inflammation is a defense against infection.

4. The most active phagocytic cells are __________________ and _______________.
5. Monocytes give rise to ___________________________.

IX. Specific Defenses (pp. 632–47)

A. Answer these questions concerning immunity and the origin of lymphocytes.

1. What is immunity?

2. A foreign substance to which lymphocytes respond is called a(n) ______________.

3. Describe a hapten. List some substances that act as haptens.

4. Where do lymphocytes originate?

5. Why are some lymphocytes called T cells while others are called B cells?

B. Answer the following concerning the functions of T cells and cell mediated immunity.

1. What are the functions of T cells?

2. How do T cells carry out these functions?

3. T cells are activated by ______________ ______________ ______________.

4. Describe the function of:

   Helper T cells

   Memory T cells

   Cytotoxic T cells

   Natural killer cells

5. Define cell-mediated immunity and antibody-mediated immunity.

6. Describe a B cell clone and explain its function.

7. Describe the roles of monoclonal antibodies and cytokines in immunotherapy.

C. Fill in the following chart.

Types of immunoglobulins		
Immunoglobulin	**Occurrence**	**Function**
IgG		
IgA		
IgM		
IgD		
IgE		

D. Describe the following antibody actions.

agglutination

precipitation

neutralization

lysis

complement

chemotaxis

opsonization

inflammation

E. Answer the following concerning primary and secondary immunity.

1. Define *primary immune response.*

2. Define *secondary immune response.*

F. Define these terms concerning types of immunity.

naturally acquired active immunity

artificially acquired active immunity

artificially acquired passive immunity

naturally acquired passive immunity

G. Describe the events of an allergic reaction.

H. Answer these questions concerning transplantation and tissue rejection.

1. What is a tissue rejection reaction?

2. In what two ways can this reaction be minimized?

I. Answer the following questions about autoimmune diseases.

1. What are autoimmune diseases?

2. How is the expression of a specific autoimmune disease controlled?

J. Describe the pathologic events of acquired immune deficiency syndrome (AIDS).

K. Answer the following concerning AIDS.

1. Trace the spread of AIDS since 1981.

2. What is the current treatment of AIDS? How successful is this treatment?

X. Clinical Focus Question

Your young adult son or daughter is leaving for college and you are concerned about reports of the increasing incidence of AIDS on college campuses. What issues would you discuss with your son or daughter? Remember to consider your personal values about sexuality and human dignity.

When you have completed the study activities to your satisfaction, retake the mastery test and compare your performance with your initial attempt. If there are still areas you do not understand, repeat the appropriate study activities.

CHAPTER 17
DIGESTIVE SYSTEM

OVERVIEW

This chapter is about the digestive system, which processes food so that nutrients can be absorbed and utilized by the cells. It names the organs of the digestive system and describes their locations and functions (objectives 1 and 2). The structure, movement, and digestive mechanisms of the alimentary canal are explained (objectives 3, 4, 7, and 8). The function and regulation of the secretions of the digestive organs are introduced (objectives 5 and 6). This chapter also tells how nutrients are absorbed from the alimentary canal (objective 9).

Studying the digestive system will help you to understand how fuel is made available for metabolism.

CHAPTER OBJECTIVES

After you have studied this chapter, you should be able to:

1. Name and describe the locations and major parts of the organs of the digestive system.
2. Describe the general functions of each digestive organ.
3. Describe the structure of the wall of the alimentary canal.
4. Explain how the contents of the alimentary canal are mixed and moved.
5. List the enzymes the digestive organs and glands secrete and describe the function of each.
6. Describe how digestive secretions are regulated.
7. Explain how digestive reflexes control movement of material through the alimentary canal.
8. Describe the mechanisms of swallowing, vomiting, and defecating.
9. Explain how the products of digestion are absorbed.

FOCUS QUESTION

How does the body make the nutrients in a ham sandwich available for utilization by the cells?

MASTERY TEST

Now take the mastery test. Do not guess. Some questions may have more than one correct answer. As soon as you complete the test, correct it. Note your successes and failures so that you can read the chapter to meet your learning needs.

1. Evidence indicates that the most frequent cause of ulcers is
   a. stress.
   b. increased acid production.
   c. increased fragility of the gastric mucosa.
   d. a bacterial infection.
2. The mechanical and chemical breakdown of food into forms that can be absorbed by cell membranes is ________________.
3. The mouth, pharynx, esophagus, stomach, small intestine, and large intestine make up the ______________ ______________ of the digestive system.
4. The salivary glands, liver, gallbladder, and pancreas are considered ______________ ______________.
5. The vessels that nourish the structures of the alimentary canal are found in the
   a. mucous membrane.
   b. submucosa.
   c. muscular layer.
   d. serous layer.
6. The two basic types of movement of the alimentary canal are ______________ movements and ______________ movements.

7. Does statement *a* explain statement *b?* ______________

a. Peristalsis is stimulated by stretching the alimentary tube.

b. Peristalsis acts to move food along the alimentary canal.

8. When the alimentary canal is being stimulated by the parasympathetic division

a. the impulses are arising in the brain and the thoracic segment of the spinal cord.

b. nerve impulses are conducted along the vagus nerve to the esophagus, stomach, pancreas, gallbladder, small intestine, and parts of the large intestine.

c. the activity of the organs of the digestive system is increased.

d. sphincter tone in the anus is increased.

9. Which of the following is/are function(s) of the mouth?

a. speech

b. beginning digestion of protein

c. pleasure

d. altering the size of pieces of food

10. The tongue is anchored to the floor of the mouth by a fold of membrane called the ______________.

11. During swallowing, muscles draw the soft palate and uvula upward to

a. move food into the esophagus.

b. enlarge the area to accommodate a bolus of food.

c. separate the oral and nasal cavities.

d. move the uvula from the path of the food bolus.

12. The third set of molars is sometimes called the ______________ ______________.

13. The teeth that bite off pieces of food are

a. incisors.

b. bicuspids.

c. canines.

d. molars.

14. The material that covers the crown of the teeth is

a. cementum.

b. dentin.

c. enamel.

d. plaque.

15. Which of the following is *not* a function of saliva?

a. cleansing of mouth and teeth

b. dissolving chemicals necessary for tasting food

c. helping in formation of food bolus

d. beginning digestion of protein

16. Stimulation of salivary glands by parasympathetic nerves will (increase/decrease) the production of saliva.

17. The salivary glands that secrete amylase are the ______________ glands.

a. submaxillary

b. parotid

c. sublingual

18. The eustachian tubes connect the middle ear and the

a. oral pharynx.

b. nasal pharynx.

c. buccal cavity.

d. oral cavity.

19. When food enters the esophagus, it is transported to the stomach by a movement called ______________.

20. When stomach contents flow back into the esophagus, they irritate the esophagus and cause a pain called ______________.

21. The area of the stomach that acts as a temporary storage area is the ______________ region.

a. cardiac

b. fundic

c. body

d. pyloric

22. Forceful vomiting after feeding in a newborn is a sign of ______________ ______________ ______________.

23. The chief cells of the gastric glands secrete

a. mucus.

b. hydrochloric acid.

c. digestive enzymes.

d. potassium chloride.

24. The digestive enzyme pepsin secreted by gastric glands begins the digestion of

a. carbohydrate.
b. protein.
c. fat.

25. The intrinsic factor secreted by the stomach aids in the absorption of ______________ from the small intestine.

26. The release of the hormone somatostatin (increases/decreases) the release of hydrochloric acid by parietal cells.

27. The release of gastrin is stimulated by

a. the sympathetic nervous system.
b. the parasympathetic nervous system.
c. histamine.
d. somatostatin.

28. The presence of food in the small intestine eventually (inhibits/increases) gastric secretion.

29. The semifluid paste formed in the stomach by mixing food and gastric contents is ______________.

30. The foods that stay in the stomach the longest are high in

a. fat.
b. protein.
c. carbohydrate.

31. The enterogastric reflex (stimulates/inhibits) peristalsis.

32. Pancreatic enzymes travel along the pancreatic duct and empty into the

a. duodenum.
b. jejunum.
c. ileum.

33. Which of the following enzymes is present in secretions of the mouth, stomach, and pancreas?

a. amylase
b. lipase
c. trypsin
d. lactase

34. Which of the following is secreted by the pancreas in an inactive form and is activated by a duodenal enzyme?

a. nuclease
b. trypsin
c. chymotrypsin
d. carboxypeptidase

35. The secretions of the pancreas are (acidic/alkaline).

36. The thick mucus that destroys lung and pancreatic tissue in victims of cytic fibrosis is the result of

a. abnormal mucus production that causes increases in mucus.
b. a defect in mucous membrane structure.
c. defective chloride channels in the cell membranes of many tissues.
d. a deficiency of pancreatic enzyme production.

37. The liver is located in the ______________ ______________ quadrant of the abdomen.

38. The most vital liver functions are those that are related to metabolism of

a. carbohydrates.
b. fats.
c. proteins.

39. Extra iron is stored by the liver in the form of ______________.

40. Nutrients are brought to the liver cells via the

a. central vein.
b. liver capillaries.
c. hepatic sinusoids.
d. connective tissue of the lobes of the liver.

41. The type(s) of hepatitis that are blood borne is(are)

a. hepatitis A.
b. hepatitis B.
c. hepatitis C.
d. hepatitis D.
e. hepatitis E.

42. The function of the gallbladder is to ______________ bile.

a. store
b. secrete
c. activate
d. concentrate

43. The hepatopancreatic sphincter is located between the

a. pancreatic duct and the common bile duct.
b. hepatic duct and the cystic duct.
c. common bile duct and the duodenum.
d. pancreatic duct and the duodenum.

44. Which of the following is/are the function(s) of bile?

a. emulsification of fat globules
b. absorption of fats
c. increase the solubility of amino acids
d. absorption of fat-soluble vitamins

45. List the portions of the small intestine: _______________, _______________, _______________.

46. The velvety appearance of the lining of the small intestine is due to the presence of

a. cilia.
b. villi.
c. mucus secreted by the small intestine.
d. capillaries.

47. The intestinal enzyme that breaks down fats is

a. sucrase.
b. maltase.
c. lipase.
d. intestinal amylase.

48. Lactose intolerance leads to an inability to digest

a. red meat.
b. eggs.
c. leafy vegetables.
d. milk and dairy products.

49. Which of the following transport mechanisms is/are not used by the small intestine?

a. diffusion
b. osmosis
c. filtration
d. active transport

50. Diarrhea results from an intestinal movement called _______________ _______________.

51. The small intestine joins the large intestine at the _______________.

52. The only significant secretion of the large intestine is

a. potassium.
b. mucus.
c. chyme.
d. water.

53. The only nutrients normally absorbed in the large intestine are _______________ and _______________.

54. Which of the following are synthesized by the bacteria of the colon?

a. gas
b. electrolytes
c. vitamin K
d. ascorbic acid

STUDY ACTIVITIES

I. Definition of Key Terms

Define the following key terms used in this chapter.

absorption

accessory organ

alimentary canal

bile

chyme

circular muscle

deciduous

feces

gastric juice

intestinal juice

intrinsic

longitudinal muscle

mesentery

mucous membrane

pancreatic juice

peristalsis

serous layer

sphincter muscle

villus

II. Introduction and General Characteristics of the Alimentary Canal (pp. 656–58)

A. Describe the method used by Marshall and Warren to demonstrate the bacterial cause of ulcers.

B. What is the name of the ulcer-producing bacterium?

C. How is the presence of H. pylori diagnosed?

D. Define *digestion.*

E. Label the following on the diagram of the digestive system: mouth, tongue, tooth, sublingual salivary gland, esophagus, liver, gallbladder, duodenum (of small intestine), large intestine, small intestine, anal canal, rectum, pancreas, stomach, pharynx, submandibular salivary gland, parotid salivary gland.

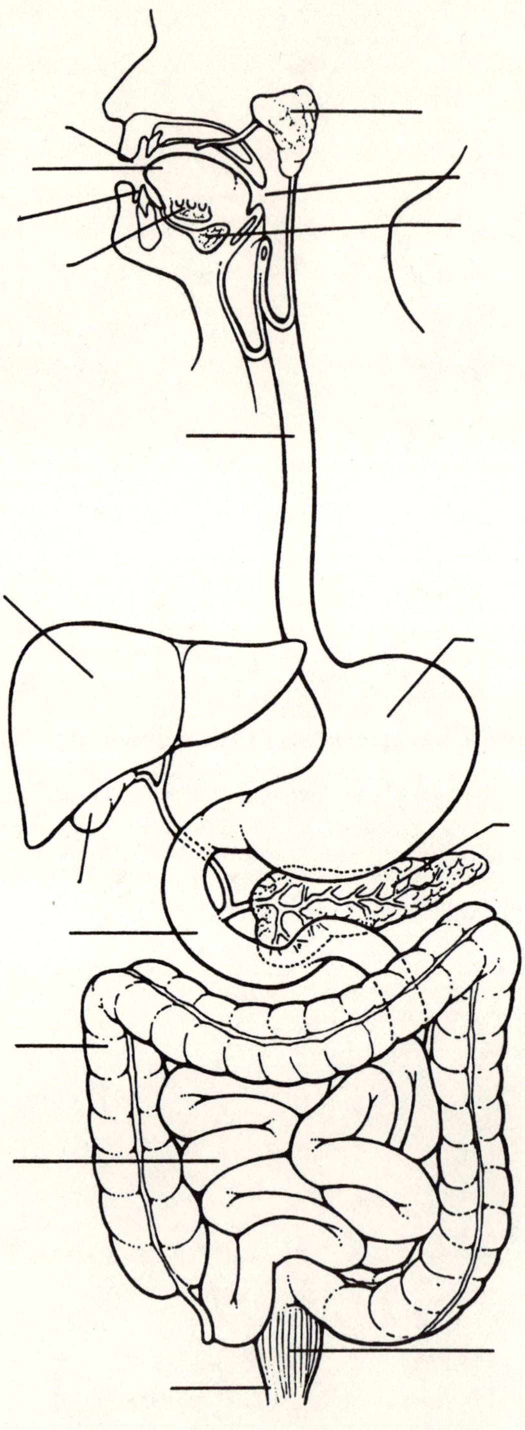

F. Fill in the following chart.

Structure and function of the alimentary tube	Composition	Function
Mucous membrane		
Submucosa		
Muscular layer		
Serous layer		

G. The two types of movement of the alimentary tube are _______________ and _______________.

H. Parasympathetic stimulation provokes a(n) _______________ in the activity of the tube; sympathetic stimulation provokes a(n) _______________ in the activity of the tube.

III. Mouth (pp. 658–64)

A. What are the functions of the mouth?

B. Answer the following concerning the cheeks and lips.

1. Describe the structure of the cheeks.

2. Describe the structure of the lips.

3. Why are cheek cells used to identify carriers of cystic fibrosis?

C. Answer the following concerning the tongue.

1. Describe the muscular structure of the tongue.

2. What is the function of the tongue?

D. Answer these questions concerning the palate.

1. What are the parts of the palate?

2. What is the function of the palate?

3. Why is infection of the palatine tonsils significant?

4. Where are the adenoids located?

E. Answer the following concerning the teeth.

1. Use yourself or a partner to locate the central incisors, lateral incisors, cuspids, first bicuspids, second bicuspids, first, second, and third molars. Are the wisdom teeth present?

2. Describe the development of teeth.

3. In the following illustration, label the crown, root, enamel, dentin, gingiva, pulp cavity, cementum, alveolar process, periodontal ligament, and root canal.

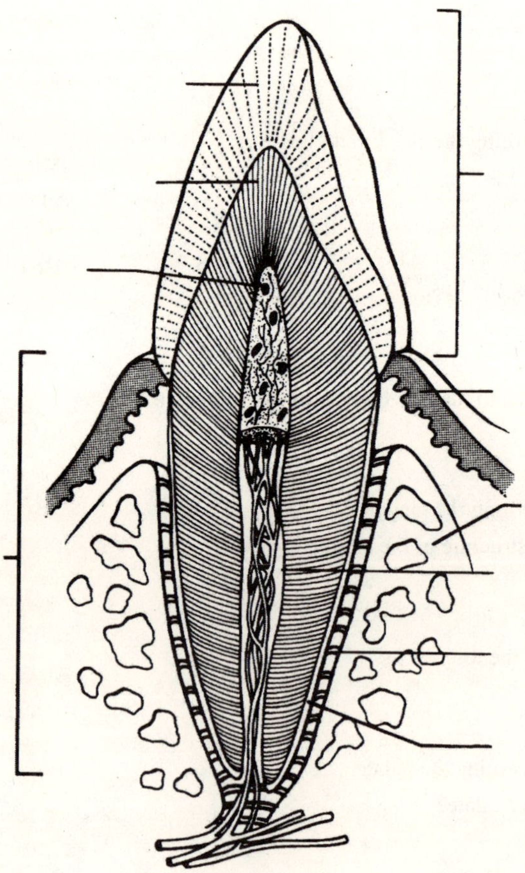

4. Identify the functions of the following kinds of teeth.
   incisors

   cuspids (canines)

   bicuspids and molars

5. What are dental caries and endodontitis? How can these conditions be prevented?

IV. Salivary Glands (pp. 665–66)

A. Answer these questions concerning salivary glands and their secretions.
   1. What is the function of salivary glands?

   2. What stimulates them to secrete saliva?

B. Fill in the following chart.

Salivary glands

Gland	Location of the gland	Secretion	Location of the duct
Parotid			
Submandibular			
Submaxillary			
Sublingual			

V. Pharynx and Esophagus (pp. 666–69)

A. Describe the nasopharynx, oropharynx, and laryngopharynx. In the description, include the muscles in the wall of the larynx.

B. List the events of swallowing.

C. Describe the structure and functions of the esophagus.

D. The esophagus passes through the diaphragm via an opening called the ________________ ______________.

E. What is a hiatal hernia?

VI. Stomach (pp. 670–76)

A. Answer the following concerning the stomach and its parts.

1. What are the functions of the stomach?

2. Label the indicated regions of the stomach and identify the function of each region: cardiac region, fundic region, body, pyloric region, pyloric canal, duodenum, pyloric sphincter, esophagus, rugae.

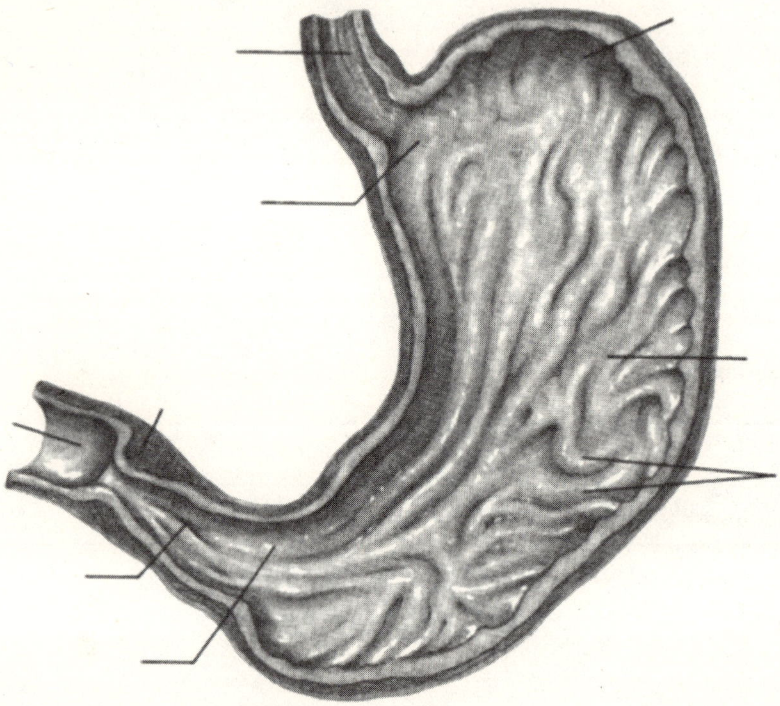

B. Describe the mucous membrane of the stomach.

C. Fill in the following chart.

Secretions of gastric glands

Cell type	Secretions	Function and action
Mucous cell		
Chief cell		
Parietal cell		

D. How are gastric secretions regulated?

E. Describe the three phases of gastric secretion.

F. What is the "alkaline tide"?

G. How can indigestion be avoided?

H. Answer the following concerning filling and emptying actions.

1. What is chyme, and how is it produced?

2. What factors affect the rate at which the stomach empties?

3. Describe vomiting.

VII. **Pancreas** (pp. 676–78)

A. Where is the pancreas located?

B. Answer the following concerning pancreatic juice.

1. Describe the action of the following pancreatic enzymes
   amylase

   lipase

   trypsin

   chymotrypsin

   carboxypeptidase

2. Describe acute pancreatitis.

3. What substance makes the pancreatic juice alkaline?

4. Describe the neural mechanisms that regulate pancreatic secretion.

5. How does secretin affect pancreatic juice?

6. How does cholecystokinin affect pancreatic juice?

7. How does cystic fibrosis affect the pancreas?

VIII. Liver (pp. 678–85)

A. Answer the following concerning the liver.

1. Describe the location and structure of the liver. Include the circulation to the liver.

2. What is the role of the liver in carbohydrate metabolism?

3. What is the role of the liver in fat metabolism?

4. What is the role of the liver in protein metabolism?

5. How does hepatic coma develop?

6. What is the role of the liver in iron homeostasis?

7. What is the digestive function of the liver?

B. Answer the following concerning bile.

1. Describe the composition of bile.

2. Which of the substances in bile is active in the digestive process?

3. What is the source of bile pigments?

C. Describe the various types of hepatitis, their modes of transmission, and their treatment.

D. Answer the following concerning the gallbladder.

1. In the following illustration, label these structures: gallbladder, hepatic ducts, cystic duct, common bile duct, pancreatic duct, hepatopancreatic sphincter, duodenum, common hepatic duct.

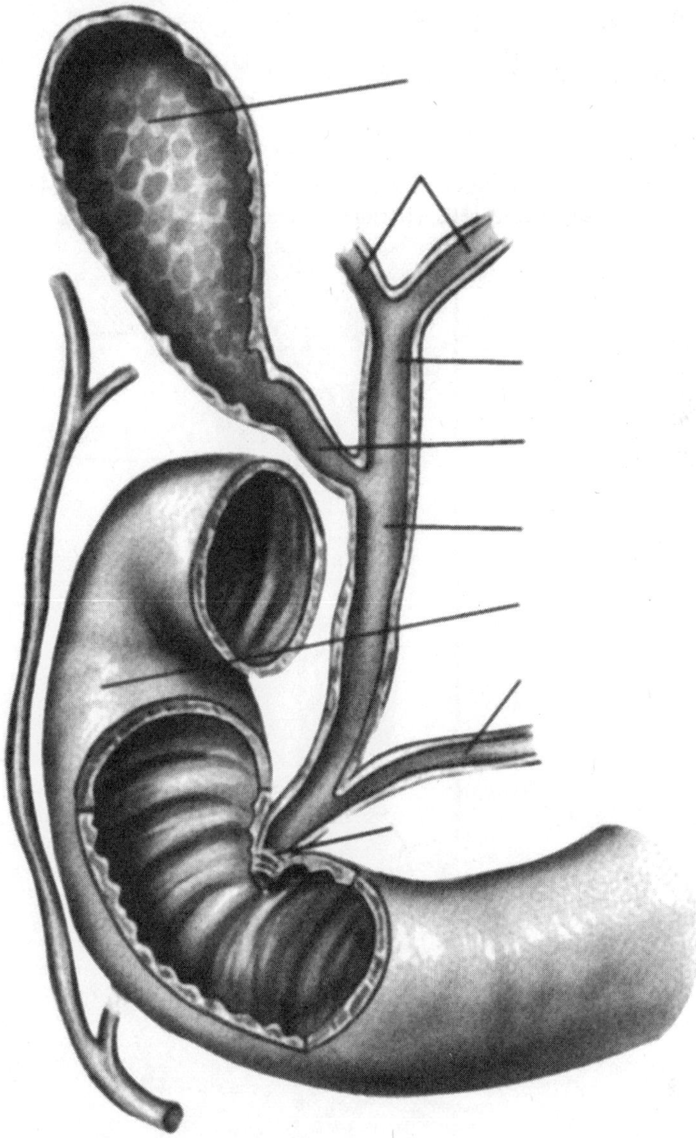

2. How is bile stored?

3. How are gallstones formed?

4. How do gallstones affect the function of the gallbladder?

E. Describe how bile is released.

F. Answer the following concerning the digestive functions of bile salts.

1. The digestive function of bile is the _______________ of bile salts.
2. Bile salts aid the absorption of _______________, _______________, and _______________.

IX. Small Intestine (pp. 685–92)

A. Name and locate the three portions of the small intestine.

B. Label the structures in the following illustration and explain the function of each: simple columnar epithelium, lacteal, capillary network, intestinal gland, goblet cells, arteriole, venule, lymph vessel, villus.

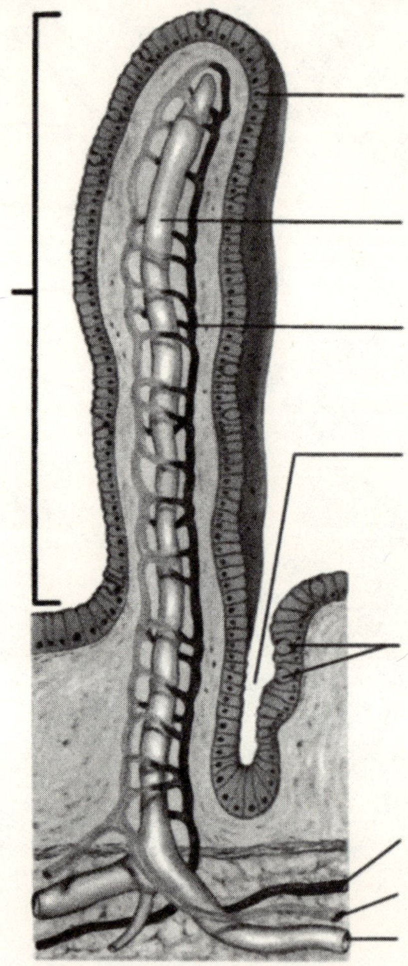

C. Describe the mechanisms that regulate intestinal secretions.

D. Answer the following concerning absorption in the small intestine.

1. Fill in the following chart.

Digestion and absorption of nutrients in the small intestine

Food	Intestinal enzyme	Nutrient	Absorption mechanism	Means of transport
Carbohydrate				
Protein				
Fat				
Electrolytes				
Water				

2. What ions are readily absorbed?

3. What ions are poorly absorbed?

4. Describe lactose intolerance and malabsorption syndrome.

E. Answer the following concerning the movements of the small intestine.

1. The normal movements of the small intestine are _______________, _______________, and _______________ movements.

2. How are these movements regulated?

3. The result of peristaltic rush is _______________.

4. What separates the small and large intestine? How does this structure work?

X. Large Intestine (pp. 692–97)

A. Label the parts of the large intestine on the following drawing: ileum; vermiform appendix; cecum; orifice of appendix; ascending, transverse, and descending colons; ileocecal sphincter; sigmoid colon; rectum; hepatic flexure; splenic flexure; tenia coli; epiploic appendage; haustra; anal canal; serous layer; muscular layer; mucous membrane.

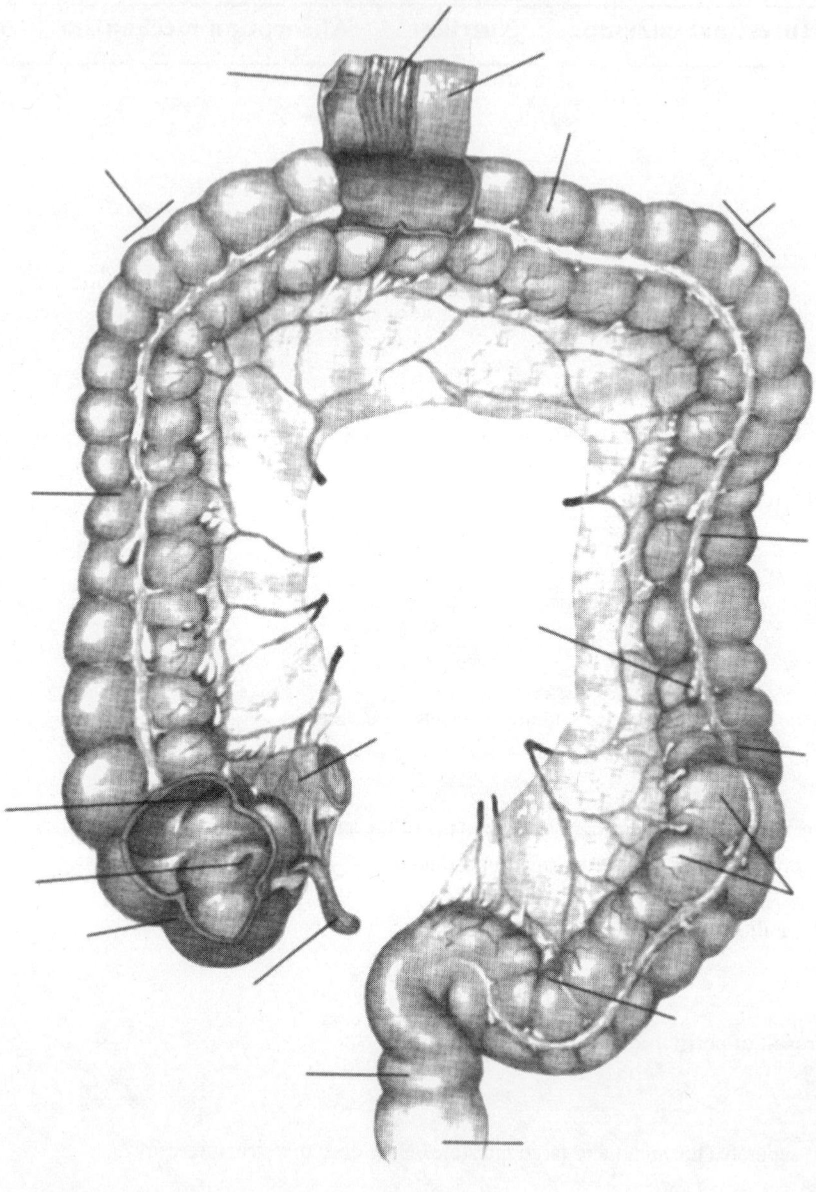

B. Answer these questions concerning the functions of the large intestine.

1. How is the rate of mucus secretion controlled?

2. What is the function of mucus?

3. What substances are absorbed in the large intestine, and what transport mechanisms are used?

4. What is the function of the bacteria in the colon?

5. Aspirin seems to reduce the death rate from cancers of the GI tract by up to 40%. Discuss the alternative explanations for this phenomenon.

C. Answer the following concerning the movements of the large intestine.

1. Describe the movements of the large intestine.

2. List the events of the defecation reflex.

D. What is the composition of feces?

E. Discuss the following disorders of the large intestine.

diarrhea

constipation

diverticulosis

ulcerative colitis

colon cancer

XI. Clinical Focus Question

Your grandmother is 80 years old, and both she and your grandfather describe themselves as being in excellent health. Your grandmother runs the household for the two of them as she has always done, including preparing all the meals. Your grandmother tells you that both your grandparents have had some bowel "irregularity" lately and asks you to recommend a laxative. Based on your knowledge of the alimentary tube, what advice would you give her?

When you have completed the study activities to your satisfaction, retake the mastery test and compare your performance with your initial attempt. If there are still areas you do not understand, repeat the appropriate study activities.

CHAPTER 18
NUTRITION AND METABOLISM

OVERVIEW

This chapter presents some basic concepts of nutrition. It defines nutrition and primary and secondary malnutrition (objectives 1 and 17). It describes the utilization, function, and sources of proteins, carbohydrates, lipids, vitamins, and minerals (objectives 2-5 and 11-15). It explains nitrogen balance, energy, energy value of foods, basal metabolic rate, adequate diet, and desirable weight (objectives 6-10, 16, and 17).

Knowledge of what foods to select and in what quantities provides a firm foundation for the study of nutrition.

CHAPTER OBJECTIVES

After you have studied this chapter, you should be able to:

1. Define *nutrition, nutrients,* and *essential nutrients.*
2. List the major sources of carbohydrates, lipids, and proteins.
3. Describe how cells utilize carbohydrates.
4. Describe how cells utilize lipids.
5. Describe how cells utilize amino acids.
6. Define *nitrogen balance.*
7. Explain how the energy values of foods are determined.
8. Explain the factors that affect an individual's energy requirements.
9. Define *energy balance.*
10. Explain what is meant by *desirable weight.*
11. List the fat-soluble and water-soluble vitamins.
12. Describe the general functions of each vitamin.
13. Distinguish between a vitamin and a mineral.
14. List the major minerals and trace elements.
15. Describe the general functions of each mineral and trace element.
16. Describe an adequate diet.
17. Distinguish between primary and secondary malnutrition.

FOCUS QUESTION

What criteria can be used to select a diet that will meet but not exceed the body's requirements for energy, growth, and repair?

MASTERY TEST

Now take the mastery test. Do not guess. Some questions may have more than one correct answer. As soon as you complete the test, correct it. Note your successes and failures so that you can read the chapter to meet your learning needs.

1. A disease found primarily in young women in which the woman deprives herself of food because she perceives herself as fat is _______________ _______________.
2. List the macronutrients.

3. List the micronutrients.

4. Nutrients, such as amino acids and fatty acids, that are necessary for health but cannot be synthesized in adequate amounts by the body are called _______________ _______________.

5. Neuropeptide Y links messages concerning nutrient use with regulators of food intake in the
   a. cerebral cortex.
   b. acurate and paraventricular nuclei.
   c. thalamus.
   d. pituitary gland.

6. Carbohydrates are ingested in such foods as
   a. meat and seafood.
   b. bread and pasta.
   c. butter and margarine.
   d. bacon.

7. A carbohydrate that cannot be broken down by human digestive enzymes and that facilitates muscle activity in the alimentary tube is _______________.

8. Energy is released from glucose in a process called _______________.

9. Glucose can be stored as glycogen in the
   a. blood plasma.
   b. muscles.
   c. connective tissue.
   d. liver.

10. The organ most dependent on an uninterrupted supply of glucose is the
    a. heart muscle.
    b. liver.
    c. adrenal gland.
    d. brain.

11. The estimated daily amount of carbohydrate needed to avoid utilization of fats and proteins for energy sources is
    a. 30–50 grams.
    b. 125–175 grams.
    c. 50–275 grams.
    d. 300–325 grams.

12. Circulating fats can be removed from the blood and their molecular structure changed by the _______________.

13. An essential fatty acid that cannot be synthesized by the body is _______________.

14. Which of the following lipids is/are found in bile salts?
    a. cholesterol
    b. phospholipids
    c. sterols
    d. vitamin A

15. The primary physiological function of lipids is to
    a. provide absorption of fat-soluble vitamins.
    b. conserve heat.
    c. provide material for the synthesis of hormones.
    d. supply energy.

16. A lipid that furnishes molecular components for the synthesis of sex hormones and some adrenal hormones is _______________.

17. Although the amounts and types of fats needed for optimal health are unknown, it is generally believed that the average American diet contains (too much/too little) fat.

18. Proteins function as
    a. enzymes that regulate metabolic reactions.
    b. promoters of calcium absorption.
    c. energy supplies.
    d. structural materials in cells.

19. Proteins are absorbed and transported to cells as _______________ _______________.

20. A protein that contains adequate amounts of the essential amino acids is called a _______________ protein.

21. Does statement *a* explain statement *b?* _______________
    a. Carbohydrate is used before other nutrients as an energy source.
    b. Carbohydrate has a protein-sparing effect.

22. The balance between gain and loss of protein in the body is characterized by
    a. nitrogen balance.
    b. homeostasis.
    c. optimal nutrition.

23. Protein requirements are relatively high during
    a. pregnancy.
    b. young adulthood.
    c. early childhood.
    d. old age.

24. The amount of potential energy contained in a food is expressed as _______________.

25. The rate the body at rest expends energy is called the ______________ ______________ ______________.

26. The resting body's expenditure of energy is expected to increase in which of the following situations?

a. rise in body temperature above 100°F
b. decreased levels of thyroxine
c. weight loss sufficient to affect body surface area
d. increase in room temperature

27. Growing children and pregnant women need relatively increased amounts of

a. protein.
b. calories.
c. fats.
d. carbohydrate.

28. To lose excess weight, a person must create a ______________ energy balance.

29. Tables of desirable weights are based on

a. longevity within a population.
b. estimated muscle mass.
c. average weight of people 25–30 years of age.
d. computation of bone and muscle size.

30. Vitamins A, D, E, and K are ______________ soluble.

31. Which of the following statements about vitamin A is *not* true?

a. Vitamin A is found only in foods of animal origin.
b. Children are less likely to develop vitamin A deficiency than adults.
c. Vitamin A can be synthesized by the body from carotenes.
d. Vitamin A is essential for the synthesis of rhodopsin.

32. Megadoses of vitamin A taken by pregnant women may cause birth defects.

a. True
b. False

33. Vitamin D_3 is produced by exposure of 7-dehydrocholesterol in the skin to ______________.

34. Exposure to excessive amounts of vitamin D will cause

a. rickets.
b. osteomalacia.
c. calcification of soft tissue.

35. Vitamin E is found in highest concentrations in the

a. bone.
b. muscle.
c. central nervous system.
d. pituitary and adrenal glands.

36. The vitamin essential for normal blood clotting is

a. vitamin C.
b. vitamin K.
c. vitamin E.
d. vitamin B.

37. Which of the B complex vitamins can be synthesized from tryptophan by the body?

a. niacin
b. thiamine
c. riboflavin
d. pyridoxine

38. The transport of which of the following is regulated by intrinsic factor and calcium?

a. folic acid
b. pantothenic acid
c. cyanocobalamin
d. biotin

39. A deficiency of which of the following vitamins can lead to a neural tube defect in a fetus?

a. biotin
b. ascorbic acid
c. pantothenic acid
d. folic acid

40. Which of the following is *not* a good source of vitamin C?

a. potatoes
b. tomatoes
c. grains
d. cabbage

41. The most abundant minerals in the body are ______________ and ______________.

42. Calcium is necessary for all but which of the following?

a. blood coagulation
b. muscle contraction
c. color vision
d. nerve impulse conduction

43. Which of the following is least likely to be a reason for potassium deficiency?

a. diarrhea
b. inadequate intake of potassium
c. vomiting
d. diuretic drugs

44. The blood concentration of sodium is regulated by the kidneys, which are influenced by the hormone ______________.

45. Chlorine is usually ingested with ______________.

46. The organelle most associated with magnesium is the

a. cell membrane.
b. Golgi apparatus.
c. endoplasmic reticulum.
d. mitochondrion.

47. Iron is associated with the body s ability to transport ______________.

48. The best natural source(s) of iron is/are

a. liver.
b. red meat.
c. egg yolk.
d. raisins.

49. A substance necessary for bone development, melanin production, and myelin formation is

a. iron.
b. iodine.
c. copper.
d. zinc.

50. The type of malnutrition that is due to poor food intake is ______________ malnutrition.

51. Kwashiorkor is due to inadequate intake of

a. calories.
b. proteins.
c. carbohydrates.
d. fats.

52. Athletes should get the majority of their calories from

a. proteins.
b. carbohydrates.
c. fats.

STUDY ACTIVITIES

I. Definition of Key Terms

Define the following key terms used in this chapter.

acetyl coenzyme A

antioxidant

basal metabolic rate

calorie

calorimeter

dynamic equilibrium

energy balance

malnutrition

mineral

nitrogen balance

nutrient

triglyceride

vitamin

II. Introduction and Carbohydrates (pp. 704–6)

A. Answer the statements concerning nutritional requirements.

1. List the six nutrients needed by the body.

2. Vegetarians are at risk for inadequate intake of _______________ _______________ _______________ and _______________ unless they make careful food choices.

3. Compare digestion and metabolism.

4. Nutrients needed for health that cannot be synthesized in adequate amounts by the body are called _______________ _______________.

5. The control center for maintaining the balance between nutrient utilization and nutrient procurement is located in parts of the _____________________.

B. Answer the following concerning sources and utilization of carbohydrates.

1. Carbohydrates are _______________ compounds that are used primarily to supply _______________.

2. In what forms are carbohydrates ingested?

3. In what forms are carbohydrates absorbed?

4. What form of carbohydrates is most commonly used by the cell as fuel?

5. Identify the areas in which glucose is stored for rapid mobilization.

6. How does cellulose facilitate intestinal function?

C. Answer these questions concerning carbohydrate requirements.

1. What is the function of carbohydrates?

2. How do sugar substitutes sweeten with fewer calories?

3. What cells are particularly dependent on a continuous supply of glucose?

4. What is the estimated daily requirement for carbohydrate?

5. What is the usual daily carbohydrate intake of Americans?

III. Lipids (pp. 706–7)

A. In what forms are lipids usually ingested?

B. What are the end products of triglyceride digestion?

C. Answer the following concerning the utilization of lipids.

1. Describe the role of the liver in the utilization of lipids.

2. An essential fatty acid that cannot be synthesized by the body is _______________ _______________.
3. Describe the role of adipose tissue in the utilization of lipids.

4. What is the function of lipids?

5. Describe how the body uses cholesterol and triglycerides.

D. Answer these questions concerning lipid requirements.

1. What is the estimated daily requirement for lipids?

2. In the average American's diet, what amount of calories is supplied by fat?

E. What are some of the health issues involved in food labeling?

IV. Proteins (pp. 707–10)

A. Answer these questions concerning sources of protein.

1. In what form is protein transported and utilized by cells?

2. What is an essential amino acid? List the essential amino acids.

3. What is the difference between a complete and an incomplete protein?

4. Why must various sources of vegetable protein be combined in a meal?

B. Answer these questions concerning the utilization of amino acids.

1. How are proteins used by the body?

2. Why is adequate carbohydrate in the diet described as protein sparing?

C. An individual whose rate of protein synthesis equals his or her rate of protein breakdown is said to be in a state of _______________ _______________.

D. Answer these questions concerning protein requirements.

1. What factors influence protein requirements?

2. How is the protein requirement for a normal adult determined?

V. Energy Expenditures (pp. 710–13)

A. Answer these questions concerning the energy values of foods.

1. What are a calorie and a kilocalorie?

2. What are the energy values of a gram each of carbohydrate, protein, and fat?

B. Answer the following concerning energy requirements.

1. What is the basal metabolic rate?

2. The apparatus used to determine BMR measures the amount of _______________ taken in and the amount of _______________ _______________ given off in a resting state.

3. What factors affect the BMR?

C. When caloric intake in food equals caloric output in the form of energy, an individual is said to be in a state of ______________ ______________.

D. Answer these questions concerning desirable weight.

1. What is desirable weight?

2. How is this different from average weight?

3. The accumulation of excessive body fat is ______________.

VI. Vitamins (pp. 713–22)

A. Answer these questions concerning vitamins.

1. What is a vitamin?

2. What are the general characteristics of fat-soluble vitamins?

3. What are the general characteristics of water-soluble vitamins?

B. Fill in the following chart.

Vitamins

Vitamin	Characteristics	Functions	Sources
Vitamin A			
Vitamin D			
Vitamin E			
Vitamin K			
Vitamin B complex			
Thiamine (B_1)			
Riboflavin (B_2)			
Niacin (nicotinic acid)			
Pyridoxine (B_6)			
Pantothenic acid			

Cyanocobalamin (B_{12})

Folacin (folic acid)

Biotin

Ascorbic acid (vitamin C)

VII. **Minerals** (pp. 722–28)

A. Describe the general characteristics of minerals.

B. Fill in the following chart.

Major minerals

Mineral	Distribution	Regulatory mechanism	Functions	Sources
Calcium				
Phosphorus				
Potassium				
Sulfur				
Sodium				
Chlorine				
Magnesium				

C. Answer the following concerning trace elements.

1. What is a trace element?

2. List the trace elements, and identify the function of each.

VIII. **Healthy Eating—The Food Pyramid and Reading Labels** (pp. 728–32)

A. What is an adequate diet?

B. Describe the food pyramid.

C. Define primary and secondary malnutrition.

D. Compare marasmus and kwashiorkor.

E. Compare anorexia nervosa and bulimia.

F. Describe a healthy eating pattern for athletes.

IX. Clinical Focus Question

A close friend is planning to be married in three months and tells you that she must lose 10 pounds so that her wedding gown will fit properly. She states that she is under a great deal of stress because of the multiple things that must be done in planning the wedding. How would you advise her?

When you have completed the study activities to your satisfaction, retake the mastery test and compare your performance with your initial attempt. If there are still areas you do not understand, repeat the appropriate study activities.

CHAPTER 19
RESPIRATORY SYSTEM

OVERVIEW

The respiratory system permits the exchange of oxygen needed for cellular metabolism for carbon dioxide, which is a by-product of cellular metabolism. This chapter describes the location and function of the organs of the respiratory system and explains how they contribute to the overall function of the system (objectives 1–3). Respiratory and nonrespiratory air movements, respiratory volumes and capacities, and how normal breathing is controlled (objectives 4–9) are discussed. Gas exchange and transport, and oxygen utilization by the cell (objectives 10–13) are explained.

There are four events in respiration: (1) exchange of gas with the external environment, (2) diffusion of gas across the respiratory membrane, (3) transport of gas to and from the cell, and (4) diffusion of gas across the cell membrane for utilization of oxygen by the cell. An understanding of all these events and how they interact is basic to understanding how cells produce the energy necessary for life.

CHAPTER OBJECTIVES

After you have studied this chapter, you should be able to:

1. List the general functions of the respiratory system.
2. Name and describe the locations of the organs of the respiratory system.
3. Describe the functions of each organ of the respiratory system.
4. Explain how inspiration and expiration are accomplished.
5. Name and define each of the respiratory air volumes and capacities.
6. Explain how the alveolar ventilation rate is calculated.
7. List several nonrespiratory air movements and explain how each occurs.
8. Locate the respiratory center and explain how it controls normal breathing.
9. Discuss how various factors affect the respiratory center.
10. Describe the structure and function of the respiratory membrane.
11. Explain how the blood transports oxygen and carbon dioxide.
12. Review the major events of cellular respiration.
13. Explain how cells use oxygen.

FOCUS QUESTION

As you are taking laundry out of the dryer, you start to sneeze as dust from the dryer hits you in the face. The phone rings in the kitchen and you run up a flight of stairs to answer it. How does the respiratory system extract oxygen from the atmosphere and give off waste gases from the body so that your body can adjust to such different levels of activity?

MASTERY TEST

Now take the mastery test. Do not guess. Some questions may have more than one correct answer. As soon as you complete the test, correct it. Note your successes and failures so that you can read the chapter to meet your learning needs.

1. Carbon monoxide poisoning can be mistaken for heart disease.
   a. True
   b. False

2. Exchange of carbon dioxide and oxygen by cells is part of

a. ventilation.
b. breathing.
c. cellular respiration.
d. transport of gases.

3. The gas exchange made possible by respiration enables cells to harness the ______________ in food molecules.

4. Which of the following organs is/are part of the upper respiratory tract?

a. lungs
b. pharynx
c. bronchi
d. larynx

5. Match the functions in the first column with the appropriate part of the nose in the second column.

____ 1. warm incoming air
____ 2. trap particulate matter in the air
____ 3. prevent infection
____ 4. moisten air
____ 5. move nasal secretions to pharynx

a. mucous membrane
b. mucus
c. cilia

6. Which of the following is(are) the result of cigarette smoking?

a. paralysis of respiratory cilia
b. production of increased amounts of mucus
c. easier access to respiratory tissue by pathogenic organisms
d. loss of elasticity in the walls of respiratory passages

7. Does statement *a* explain statement *b?* ______________

a. The sinuses are air-filled spaces in bones of the skull and face.
b. Inflammation of the nose can lead to fluid being trapped in the sinuses.

8. The pharynx is the cavity behind the mouth extending from the ______________ ______________ to the ______________.

9. The portions of the larynx concerned with preventing foreign objects from entering the trachea are the

a. arytenoid cartilages.
b. glottis.
c. epiglottis.
d. hyoid bone.

10. The portion of the larynx visible in the neck as the Adam's apple is the ______________ ______________.

11. The pitch of the voice is controlled by

a. changing the tension of the vocal cords.
b. changing the force of the air passing through the larynx.
c. opening the vocal cords.
d. increasing the volume of air passing through the larynx.

12. The trachea is maintained in an open position by

a. cartilaginous rings.
b. the amount of collagen in the wall of the trachea.
c. the tone of smooth muscle in the wall of the trachea.
d. the continuous flow of air through the trachea.

13. A temporary opening in the trachea made to bypass an obstruction is a ______________.

14. The right and left bronchi arise from the trachea at the

a. suprasternal notch.
b. manubrium of the sternum.
c. fifth thoracic vertebra.
d. eighth intercostal space.

15. The smallest branches of the bronchial tree are the ______________ ______________.

16. As the lumen of the branches of the bronchial tree decreases, the amount of cartilage (increases/decreases).

17. The instrument used to examine the trachea and bronchial tree and to remove foreign objects aspirated into air passages is a ______________ ______________.

18. The type of epithelium found in the alveoli is

a. simple squamous.
b. ciliated columnar.
c. pseudostratified.
d. cuboidal.

19. Blood is pumped out of the body and across a semipermeable membrane that adds oxygen and removes carbon dioxide in

a. artificial respiration.
b. extracorporeal membrane oxygenation.
c. mechanical ventilation.
d. intravascular oxygenation.

20. Each lung is entered on its medial surface by a bronchus and blood vessels in a region called the _______________.

21. The _______________ lung is composed of superior, inferior, and middle lobes.

22. The serous membrane covering the lungs is the _______________ _______________.

23. The serous membrane covering the inner wall of the thoracic cavity is the _______________ _______________.

24. Inspiration occurs after the diaphragm _______________, thus (increasing/decreasing) the size of the thorax and (increasing/decreasing) the pressure within the thorax.

25. The other muscles that act to change the size of the thorax during normal respiration are the

a. sternocleidomastoids.
b. pectorals.
c. intercostals.
d. latissimus dorsi.

26. The safest way to deal with asbestos in a building is to leave it undisturbed.

a. True
b. False

27. Expansion of the lungs during inspiration is assisted by the surface tension of fluid in the _______________ cavity.

28. The surface tension of fluid in the alveoli is decreased by a secretion, _______________, that prevents collapse of the alveoli.

29. The force responsible for expiration comes mainly from

a. contraction of intercostal muscles.
b. change in the surface tension within alveoli.
c. elastic recoil of tissues in the lung and thoracic wall.
d. contraction of abdominal muscles to push the diaphragm upward.

30. The ease with which lungs can be expanded in response to pressure changes during breathing is called _______________.

31. The pressure in the pleural cavity is

a. greater than atmospheric pressure.
b. less than atmospheric pressure.
c. the same as atmospheric pressure.

32. Respiratory air volumes are measured by an instrument called a _______________.

33. The amount of air that enters and leaves the lungs during a normal, quiet respiration is the

a. vital capacity.
b. respiratory cycle.
c. total lung capacity.
d. tidal volume.

34. Respiratory volumes are used to calculate _______________ _______________.

35. The anatomic dead space is composed of the _______________, _______________, and _______________.

36. In a normal individual, the anatomic dead space and the physiologic dead space are (equal/not equal).

37. The amount of new air that reaches the alveoli and is available for gas exchange is represented by the _______________ _______________ rate.

38. Coughing, laughing, and yawning are examples of _______________ _______________ _______________.

39. Because of normal respiratory physiology, people with bronchial asthma will initially have difficulty with

a. inspiration.
b. expiration.
c. inspiration and expiration.

40. The pathologic events of emphysema include all of the following *except*

a. loss of elasticity in alveolar tissue.
b. loss of interalveolar walls, so that larger chambers form.
c. narrowing of the lumen of the bronchi.
d. loss of capillary network.

41. Normal breathing is controlled by the respiratory center located in the _______________ _______________.

42. The rate of breathing is controlled by the

a. medullary rhythmicity area.
b. apneustic area.
c. pneumotaxic area.

43. The Hering-Breuer reflexes are activated by

a. stretch receptors in bronchioles and alveoli.
b. an increase in hydrogen ions.
c. a decrease in oxygen saturation.
d. a sudden fall in blood pressure.

44. The most potent stimulus to increase respiratory rate and depth is to increase the blood concentration of _______________ _______________.

45. Hyperventilation leads to dizziness because of

a. an increase in blood pressure.
b. a decrease in heart rate.
c. generalized vasodilation.
d. a decrease in blood pH.

46. Exercise provokes an increase in respiratory rate due to

a. increased CO_2 levels.
b. generalized vasoconstriction.
c. stimulation of proprioceptors in joints.
d. stimulation of the respiratory center by the cerebral cortex.

47. A phagocyte that moves through alveolar pores is a _______________ _______________.

48. The respiratory membrane consists of a single layer of epithelial cells and basement membrane from a(n) _______________ and a(n) _______________.

49. The rate at which a gas diffuses from one area to another is determined by differences in _______________ in the two areas.

50. The pressure of each gas within a mixture of gases is known as its _______________ _______________.

51. Pneumonia, tuberculosis, and atelectasis present similar problems in that they

a. decrease the surface available for diffusion of gases.
b. obstruct the flow of air into the lungs.
c. diminish blood circulation to the lungs.
d. destroy surfactant.

52. Oxygen is transported to cells by combining with _______________.

53. Oxygen is released in greater amounts as carbon dioxide levels and temperature (increase/decrease).

54. Carbon monoxide interferes with oxygen transport by binding to _______________.

55. The largest amount of carbon dioxide is transported

a. dissolved in blood.
b. combined with hemoglobin.
c. as bicarbonate.
d. as carbonic anhydrase.

Study Activities

I. Definition of Key Terms

Define the following key terms used in this chapter.

alveolus

bronchial tree

carbaminohemoglobin

carbonic anhydrase

cellular respiration

citric acid cycle

expiration

glottis

hemoglobin

hyperventilation

inspiration

oxyhemoglobin

partial pressure

pleural cavity

respiratory center

respiratory membrane

respiratory volume

surface tension

surfactant

II. Introduction and Why We Breathe (pp. 739)

A. What is the mechanism of carbon monoxide poisoning?

B. List the events of respiration.

1. What terms are used to describe the movement of air in and out of the lungs?

2. The exchange of gases between the air in the lungs and blood is called _______________ _______________.

C. What events at the cellular level make it necessary for the body to take in oxygen and get rid of carbon dioxide?

III. Organs of the Respiratory System (pp. 740-51)

A. In the accompanying drawing, label these structures: nasal cavity, nostril, sinus, oral cavity, soft palate, hard palate, pharynx, larynx, trachea, bronchial tree, left lung, right lung, epiglottis.

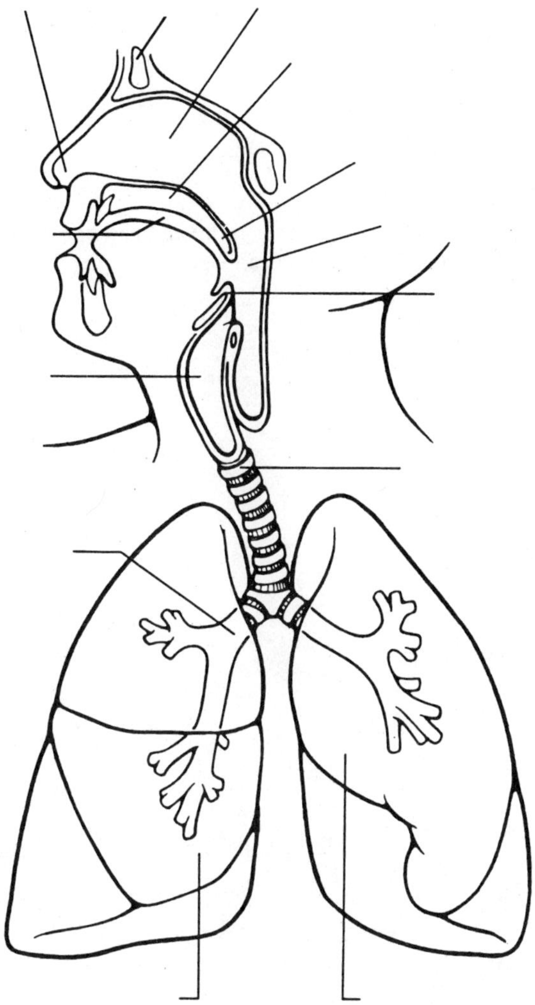

B. Match the functions in the first column with the appropriate terms in the second column.

____ 1. entrap dust

____ 2. lighten skull and provide vocal resonance

____ 3. warm and humidify air entering the nose

____ 4. provide movement to mucus layer

____ 5. bring macrophages in contact with bacteria

a. mucous membrane

b. mucus

c. paranasal sinus

d. cilia

C. How does cigarette smoking affect the respiratory system?

D. Answer the following questions about the sinuses.

1. Describe the structure and function of the paranasal sinuses.

2. Why do people experience headaches when the sinuses are inflamed?

E. What is the location and function of the pharynx?

F. Answer the following concerning the larynx.

1. Identify the structures in the accompanying drawings: hyoid bone, epiglottic cartilage, trachea, thyroid cartilage, cricoid cartilage, corniculate cartilage, arytenoid cartilage.

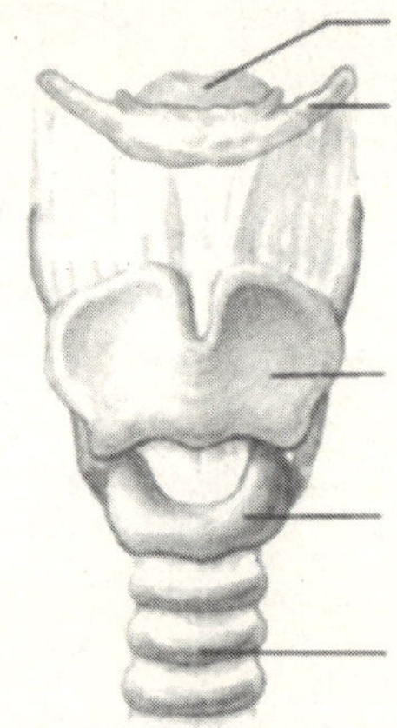

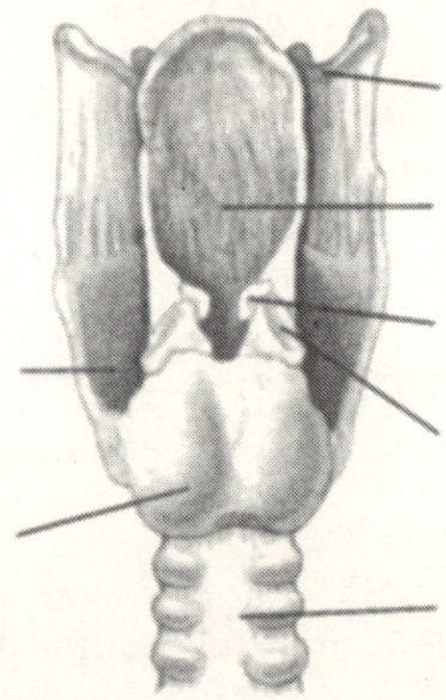

2. The structure of the larynx that helps close the glottis during swallowing is the _______________ _______________.

3. The structures of the larynx that produce sound are the _______________, _______________, and _______________.

4. The _______________ of the voice is due to the tension of the vocal cords.

5. The _______________ of the voice is due to the force of the air passing over the vocal cords.

6. Inflammation of the mucous membrane of the larynx is called _______________.

G. The trachea is prevented from collapsing by the presence of _______________ _______________ that are C-shaped.

H. What is a tracheostomy?

I. Answer the following concerning the bronchial tree.

1. On the accompanying drawing label the larynx, trachea, left superior lobe, left inferior lobe, right middle lobe, right inferior lobe, right superior lobe, terminal bronchiole, alveolar duct, right primary bronchus, secondary bronchi, tertiary bronchi, alveolus.

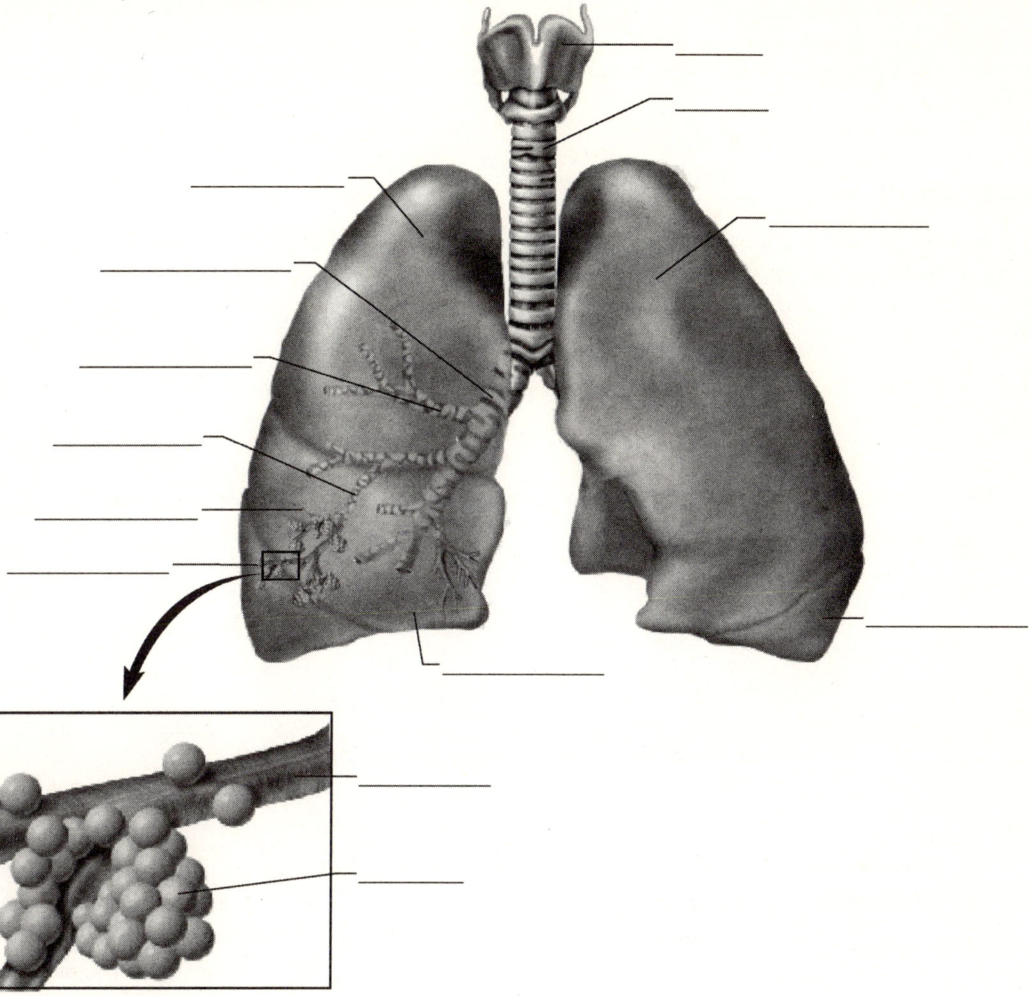

2. How does the structure of the bronchial tree change as the bronchi become smaller?

3. What is the function of alveoli?

4. Describe the structural changes that occur as the respiratory tubes become smaller.

5. The trachea and bronchial tree can be seen with an instrument called a ________________ ________________.

6. What happens to inhaled particles once they get past the trachea?

7. List some common particulate substances that cause lung disease.

J. Answer the following concerning the lungs.

1. The right lung has _______________ lobes, the left lung has _______________ lobes.
2. How is the pleural cavity formed?

3. Compare artificial respiration, extracorporeal membrane oxygenation, and use of an intravascular oxygenator.

IV. Breathing Mechanism (pp. 751-60)

A. Answer the following concerning inspiration.

1. List the events of inspiration beginning with stimulation of the phrenic nerve.

2. Describe the role of surface tension in the pleural cavity and in the alveoli.

B. Respiratory distress syndrome (hyaline membrane disease) is due to an insufficient amount of _______________.

C. What is compliance?

D. List the events of expiration.

E. What is the role of the intercostal muscles in inspiration and expiration?

F. Match the terms in the first column with the correct definition in the second column.

____ 1. inspiratory reserve volume	a. volume moved in or out of the lungs during quiet respiration
____ 2. expiratory reserve volume	b. volume that can be inhaled during forced breathing in addition to tidal volume
____ 3. residual volume	c. volume that can be exhaled in addition to tidal volume
____ 4. vital capacity	d. volume that remains in the lungs at all times
____ 5. total lung capacity	e. maximum air that can be exhaled after taking the deepest possible breath
____ 6. tidal volume	f. total volume of air the lungs can hold

G. Given the following volumes:

inspiratory reserve	2,500 cc
expiratory reserve	1,000 cc
residual	1,100 cc
tidal volume	550 cc

calculate the following lung capacities:

vital capacity

inspiratory capacity

functional residual capacity

total lung capacity

H. Describe physiologic dead space and anatomic dead space.

I. List the six nonrespiratory air movements.

J. How do each of the following disorders interfere with ventilation?
paralysis of breathing muscles

bronchial asthma

emphysema

V. **Control of Breathing** (pp. 760-64)

A. Answer the following concerning the respiratory center.

1. Label the indicated structures: vagus nerve, sensory pathway, respiratory center, motor pathways, spinal cord, rib, phrenic nerve, alveoli, intercostal nerve, external intercostal muscles, diaphragm, stretch receptors, lung.

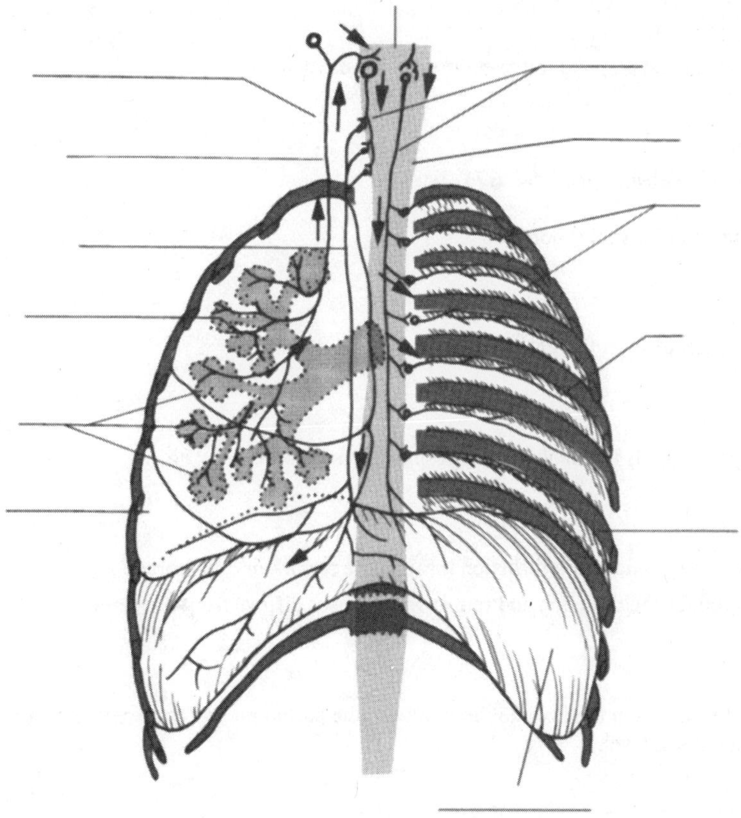

2. The area that establishes the basic rhythm of breathing is the _______________ _______________
_______________.

3. The _______________ area produces forceful, prolonged inspirations and weak expirations.

4. The _______________ area inhibits the area described in the previous statement.

B. Answer these questions concerning the factors affecting breathing.

1. What chemical factors affect breathing?

2. How do chemoreceptors in the carotid and aortic bodies affect breathing?

3. Compare sleep apnea in infants and adults.

4. Where are the receptors for the Hering-Breuer reflexes?

5. How do these reflexes affect respiration?

C. Answer these questions concerning exercise and breathing rate.

1. What factors influence the breathing rate during exercise?

2. What systems must respond to the cells' need for oxygen?

VI. Alveolar Gas Exchanges (pp. 764-67)

A. What is the function of alveolar pores?

B. What is the function of alveolar macrophages?

C. Identify the layers of cells that make up the respiratory membrane.

D. Answer the following concerning diffusion through the respiratory membrane.

1. What determines the direction and rate at which gases diffuse from one area to another?

2. Use the information in the textbook to work out the partial pressures of oxygen, carbon dioxide, and nitrogen in ordinary air at sea level.

3. Describe the diffusion of oxygen and carbon dioxide in the lungs.

4. Describe the effects of pneumonia, tuberculosis, atelectasis, and adult respiratory distress syndrome on alveolar gas exchange.

5. How and why does high altitude affect gas exchange?

VII. Gas Transport (pp. 767-72)

A. Answer the following concerning oxygen transport.

1. Describe how oxygen is transported to cells.

2. Why is oxygen released at the cell?

3. What factors affect how much oxygen is released from hemoglobin to tissues?

B. A gas that interferes with oxygen transport by forming a stable bond with hemoglobin is _______________ _______________.

C. Answer the following concerning carbon dioxide transport.

1. How is carbon dioxide transported away from the cell?

2. The diffusion of bicarbonate ions out of red blood cells into plasma and the diffusion of chloride ions from plasma into red blood cells is called _______________ _______________.

3. Why is carbon dioxide from the blood lost in the lung?

VIII. Clinical Focus Question

George L. is a 60-year-old carpenter. He smoked two packs of cigarettes a day until one year ago when he stopped smoking. Following a hospitalization for pneumonia, George was told he had emphysema.

A. How will George L.'s emphysema affect his lifestyle?

B. What might you suggest to help him cope with these changes?

C. Is the smoking history of George's family of interest to you? Why or why not?

When you have completed the study activities to your satisfaction, retake the mastery test and compare your performance with your initial attempt. If there are still areas you do not understand, repeat the appropriate study activities.

CHAPTER 20
URINARY SYSTEM

OVERVIEW

This chapter is about the urinary system, which plays a vital role in maintaining the internal environment by excreting nitrogenous waste products and by selectively excreting or retaining water and electrolytes. It identifies, locates, and describes the functions of the organs of the urinary system (objectives 1–3). It explains the structure and function of the nephron—the basic unit of function of the kidney (objective 5). It traces the pathway of blood through the blood vessels of the kidney, explains how glomerular filtrate is produced, and discusses the role of tubular reabsorption in urine production and the cause of change in the osmotic concentration of the glomerular filtrate of the renal tubule (objectives 4 and 6–10). It also describes the role of tubular secretion in urine formation (objective 11). It discusses the structure of the ureters, urinary bladder, and urethra, and how they function in micturition (objectives 12 and 13).

A study of the urinary system is basic to understanding how the body maintains its chemistry within very narrow limits.

CHAPTER OBJECTIVES

After you have studied this chapter, you should be able to:

1. Name the organs of the urinary system and list their general functions.
2. Describe the locations of the kidneys and the structure of a kidney.
3. List the functions of the kidneys.
4. Trace the pathway of blood through the major vessels within a kidney.
5. Describe a nephron and explain the functions of its major parts.
6. Explain how glomerular filtrate is produced and describe its composition.
7. Explain how various factors affect the rate of glomerular filtration and how this rate is regulated.
8. Discuss the role of tubular reabsorption in urine formation.
9. Explain why the osmotic concentration of the glomerular filtrate changes as it passes through a renal tubule.
10. Describe a countercurrent mechanism and explain how it helps concentrate urine.
11. Define *tubular secretion* and explain its role in urine formation.
12. Describe the structure of the ureters, urinary bladder, and urethra.
13. Discuss the process of micturition and explain how it is controlled.

FOCUS QUESTION

It is 90° and a perfect day for the beach. You spend most of the day on the beach, swimming and relaxing. You drink a half-gallon jug of lemonade in the course of the day. This evening you will be meeting friends at a local pub for sandwiches and beer. How do the kidneys help maintain the internal environment under such varied conditions?

MASTERY TEST

Now take the mastery test. Do not guess. Some questions may have more than one correct answer. As soon as you complete the test, correct it. Note your successes and failures so that you can read the chapter to meet your learning needs.

1. When the toxins produced by E. coli travel to the kidney, they damage the
   a. renal pyramids.
   b. efferent capillaries.
   c. glomerular capsule.
   d. cells of the tubules.

2. The organ(s) of the urinary system whose primary function is transport of urine is/(are) the

a. kidney.
b. urethra.
c. ureters.
d. bladder.

3. The kidneys are located

a. within the abdominal cavity.
b. between the twelfth thoracic and third lumbar vertebrae.
c. posterior to the parietal peritoneum.
d. just below the diaphragm.

4. The superior end of the ureters is expanded to form the funnel-like ______________ ______________.

5. The small elevations that project into the renal sinus from the substance of the kidney and form the sinus wall are called

a. renal pyramids.
b. the renal medulla.
c. renal calyces.
d. renal papillae.

6. A kidney tumor composed of embryonic cells is a ______________ tumor.

7. Which of the following are regulatory functions of the kidney?

a. excretion of metabolic wastes from the blood
b. control of red blood cell production
c. regulation of blood pressure
d. regulation of calcium absorption by activation of vitamin D

8. The blood supply to the nephrons is via the

a. renal artery.
b. interlobar artery.
c. arciform artery.
d. afferent arterioles.

9. The structure of the renal corpuscle consists of the

a. glomerulus.
b. Bowman's (glomerular) capsule.
c. descending loop of Henle.
d. proximal convoluted tubule.

10. Inflammation of the glomeruli is

a. nephritis.
b. glomerulonephritis.
c. pyelonephritis.

11. Blood leaves the capillary cluster of the renal capsule via the

a. afferent arteriole.
b. efferent arteriole.
c. peritubular capillary system.
d. renal vein.

12. The relatively high pressure in the glomerulus is due to

a. the small diameter of the glomerular capillaries.
b. the increased size of the smooth muscle in the venules that receive blood from the glomerulus.
c. the large volume of blood that circulates through the kidney relative to other body parts.
d. the smaller diameter of the efferent arteriole in comparison with the afferent arteriole.

13. The end product of kidney function is ______________.

14. The transport mechanism used in the glomerulus is

a. filtration.
b. osmosis.
c. active transport.
d. diffusion.

15. The fluid formed in the capillary cluster of the nephron is the same as blood plasma except for the presence of

a. glucose.
b. larger molecules of plasma protein.
c. bicarbonate ions.
d. creatinine.

16. Blood pressure affects urine formation because ______________ ______________ of the blood is necessary to the transport mechanism used in the glomerulus.

17. Sympathetic nerve impulses can be expected to (increase/decrease) glomerular filtration.

18. A significant fall in blood pressure, as during shock, causes an/a (increase/decrease) in glomerular filtration.

19. An increase in the level of plasma proteins will (increase/decrease) the filtration rate.

20. Does statement *a* explain statement *b?* ______________
   a. An increase in hydrostatic pressure causes a decrease in the glomerular filtration rate.
   b. An enlarged prostate gland tends to decrease the amount of urine produced.

21. How much fluid filters through the glomerulus in a 24-hour period?
   a. 1½ quarts
   b. 2 cups
   c. 45 gallons
   d. 5–10 quarts

22. Which of the following substances is present in glomerular filtrate but not in urine?
   a. urea
   b. sodium
   c. potassium
   d. glucose

23. Which of the following substances is transported by osmosis throughout the renal tubule?
   a. glucose
   b. water
   c. plasma proteins
   d. amino acids

24. Substances such as creatinine, lactic acid, sodium, and potassium ions are reabsorbed in the
   a. proximal convoluted tubule.
   b. distal convoluted tubule.
   c. descending limb of the loop of Henle.
   d. ascending limb of the loop of Henle.

25. Most (about 70%) of the water, sodium, and other ions in glomerular filtrate are reabsorbed in the
   a. loop of Henle
   b. proximal convoluted tubule.
   c. distal convoluted tubule.
   d. renal pyramid.

26. The reabsorption of sodium and chloride ions in the distal convoluted tubules and the collecting duct is influenced by ______________.

27. The mechanism that acts to continue sodium reabsorption from tubular fluid in the loop of Henle and at the same time causes the fluid to become hypotonic to its surroundings is the ______________ mechanism.

28. The permeability of the distal segment of the tubule to water is regulated by
   a. blood pressure.
   b. ADH.
   c. aldosterone.
   d. renin.

29. Urea is a by-product of ______________ catabolism.

30. The mechanism by which greater amounts of a substance may be excreted in urine than was filtered from the plasma in the glomerulus is
   a. tubular absorption.
   b. active transport.
   c. pinocytosis.
   d. tubular secretion.

31. Which of the following substances enter the urine using tubular secretion?
   a. lactic acid
   b. hydrogen ions
   c. amino acids
   d. potassium

32. Which of the following is *not* a normal constituent of urine?
   a. urea
   b. uric acid
   c. creatinine
   d. ketones

33. The normal output of urine for an adult in one hour is ______________ ml.
   a. 20–30
   b. 30–40
   c. 40–50
   d. 50–60

34. Tests of renal clearance of insulin and creatinine are used to provide information about
   a. renal blood flow.
   b. glomerular filtration.
   c. tubular reabsorption.
   d. tubular secretion.

35. Urine is conveyed from the kidney to the bladder via the ______________.

36. Urine moves along the ureters via

a. hydrostatic pressure.
b. gravity.
c. peristalsis.

37. The ureterorenal reflex will (increase/decrease) urine production in situations where urine flow from the kidney to the bladder is impeded.

38. The internal floor of the bladder has three openings in a triangular area called the _______________.

39. The third layer of the bladder is composed of smooth muscle fibers and is called the _______________ muscle.

a. micturition
b. detrusor
c. urinary
d. sympathetic

40. When stretch receptors in the bladder send impulses along the parasympathetic paths, the individual experiences a sensation known as _______________.

41. The usual amount of urine voided at one time is about _______________ cc.

42. Which of the following structures is under conscious control?

a. external urethral sphincter
b. internal urethral sphincter
c. bladder wall

43. In the female, the urinary meatus is located anterior to the _______________ and posterior to the _______________.

Study Activities

I. Definition of Key Terms

Define the following key terms that are used in this chapter.

afferent arteriole

autoregulation

countercurrent mechanism

detrusor muscle

efferent arteriole

glomerular filtration

glomerulus

juxtaglomerular apparatus

micturition

nephron

peritubular capillary

renal corpuscle

renal cortex

renal medulla

renal plasma threshold

renal tubule

retroperitoneal

tubular reabsorption

tubular secretion

II. Introduction (p. 779)

How does shigatoxin affect the kidneys?

III. Kidneys (pp. 779-89)

A. Identify the parts of the urinary system in the accompanying drawing: renal vein, renal artery, hilum, kidney, inferior vena cava, aorta, ureter, urinary bladder, urethra.

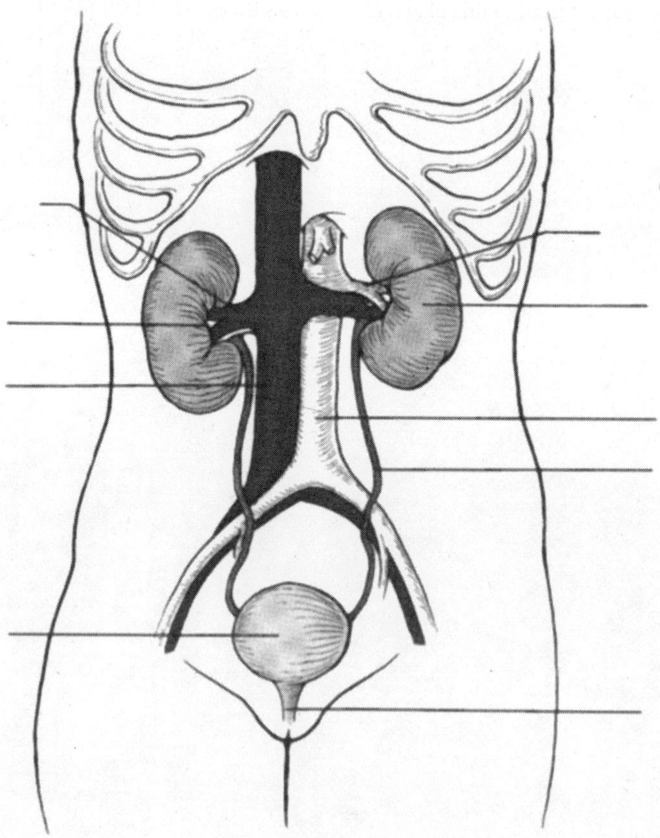

B. Describe the precise locations of the kidneys.

C. Label the structures in the accompanying drawing: renal column, renal pelvis, ureter, renal capsule, renal pyramid, major calyx, minor calyx, renal cortex, renal medulla, nephrons, renal papilla.

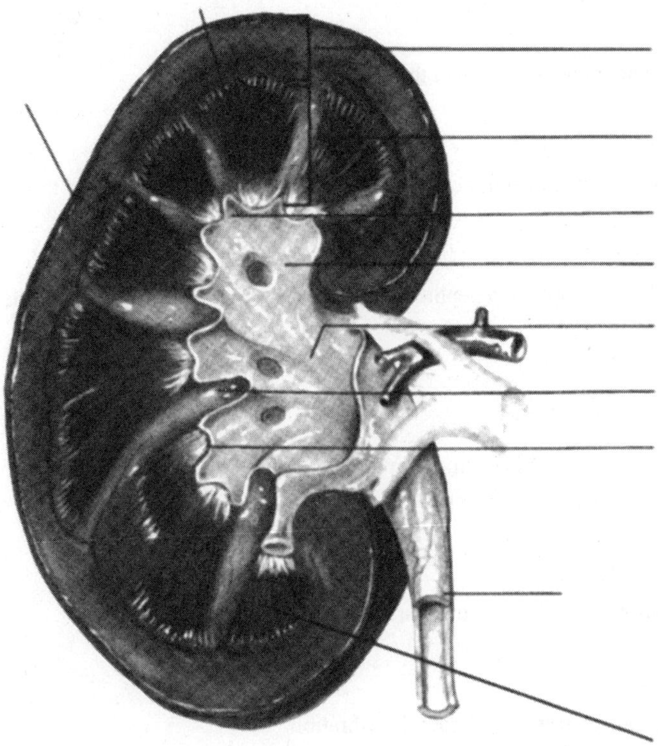

D. What is the functional unit of the kidney?

E. What are the functions of the kidney?

F. Answer the following questions about inherited kidney disorders.

1. Describe polycystic kidney disease.

2. What is a Wilms tumor and how does it develop?

G. Compare hemodialysis and continuous ambulatory peritoneal dialysis.

H. Answer the following questions about renal blood vessels.

1. From where do the renal arteries arise?

2. The renal arteries carry from _____ % to _____ % of cardiac output at rest.

3. Describe the arterial and venous pathways in the kidney.

I. Answer the following about nephrons.

1. Describe the structure of a nephron.

2. How does glomerulonephritis affect the kidney?

3. What is the role of the juxtaglomerular apparatus in regulating blood flow through the renal vessels?

4. Describe the blood supply of a nephron and state how high pressure is maintained in glomerular capillaries.

IV. Urine Formation (pp. 789-802)

A. Answer these questions concerning urine formation.

1. What are the functions of nephrons?

2. What mechanisms are used by the nephron to accomplish these functions?

B. Answer these questions concerning glomerular filtration.

1. What is glomerular filtration?

2. What is the amount and composition of glomerular filtrate?

3. What plasma substances increase when kidney function is compromised?

4. What factors affect the glomerular filtration rate? Explain your answers.

5. How does shock affect the kidneys?

6. How do the macula densa and angiotensin II regulate filtration rates?

7. What is the role of atrial natriuretic peptide in the control of blood volume?

C. Answer the following about tubular reabsorption.

1. Explain how tubular reabsorption is selective. Describe the reabsorption of glucose, amino acids, and albumin.

2. Define *renal plasma threshold.*

3. What factors can lead to glycosuria?

4. What is the nephrotic syndrome?

5. List the substances that are reabsorbed by the epithelium of the proximal convoluted tubule.

6. Describe water and sodium reabsorption in the proximal segment of the renal tubule.

7. Give a detailed description of the countercurrent mechanism.

8. Describe exactly how ADH affects the nephron.

9. How is urea reabsorbed?

D. Answer these questions about tubular secretion.

1. What is tubular secretion?

2. How are hydrogen ions secreted?

3. How are potassium ions secreted?

4. What other substances may be secreted in the tubules of the nephron?

E. Answer these questions concerning the composition of urine.

1. What is the composition of normal urine?

2. What is the normal output of urine?

3. What factors affect urinary output?

4. How does kidney function differ in children and adults?

5. What is renal clearance? How is it used to diagnose kidney disease?

6. How is urea formed and how is it excreted?

V. Elimination of Urine (pp. 802-7)

A. Answer the following about the ureters.

1. Describe the location and structure of the ureters.

2. How does the structure of the bladder and ureters promote the spread of infection?

3. How is urine moved through the ureter?

4. What happens when a ureter is obstructed by a stone?

B. Answer these questions concerning the urinary bladder.

1. How does the bladder change as it fills with urine?

2. How is urine prevented from spilling back into the ureters?

3. Describe the structure of the bladder wall.

C. Describe the process of micturition. Be sure to include both autonomic and voluntary events.

D. What is incontinence and how can it be treated?

E. What are the differences between the male and female urethra?

F. How does the composition of urine help to identify the state of an individual s health?

VI. Clinical Focus Question

Charles has chronic renal failure and undergoes dialysis three times a week. Maintenance of a stable blood pressure is a serious concern during hemodialysis. Can you explain why this is so important? How will Charles have to modify his diet to limit his intake of protein?

When you have completed the study activities to your satisfaction, retake the mastery test and compare your performance with your initial attempt. If there are still areas you do not understand, repeat the appropriate study activities.

CHAPTER 21
WATER, ELECTROLYTE, AND ACID-BASE BALANCE

OVERVIEW

This chapter presents the roles of several body systems in maintaining proper concentrations of water and electrolytes. Water and electrolyte balance is defined, and the importance of this balance is explained (objective 1). This chapter tells how water enters the body, how it is distributed in various body compartments, and how it leaves the body (objectives 2 and 3). It explains how electrolytes enter and leave the body and how their concentration is regulated (objective 4). This chapter describes acid-base balance, mechanisms regulating this balance, and the etiology and treatment of acid-base disturbances (objectives 5–9).

Water and electrolyte balance affects and is affected by the various chemical interactions of the body. A knowledge of the mechanisms that control fluid and electrolyte concentrations is essential to understanding the nature of the internal environment.

CHAPTER OBJECTIVES

After you have studied this chapter, you should be able to:

1. Explain water and electrolyte balance, and discuss the importance of this balance.
2. Describe how the body fluids are distributed within compartments, how the fluid composition differs between compartments, and how the fluids move from one compartment to another.
3. List the routes by which water enters and leaves the body, and explain how water input and output are regulated.
4. Explain how electrolytes enter and leave the body, and how the input and output of electrolytes are regulated.
5. Explain acid-base balance.
6. Describe how hydrogen ion concentrations are expressed mathematically.
7. List the major sources of hydrogen ions in the body.
8. Distinguish between strong and weak acids and bases.
9. Explain how chemical buffer systems, the respiratory center, and the kidneys minimize changing pH values of the body fluids.

FOCUS QUESTION

The actions of which body systems must be coordinated to maintain normal concentrations of fluids and electrolytes?

MASTERY TEST

Now take the mastery test. Do not guess. Some questions may have more than one correct answer. As soon as you complete the test, correct it. Note your successes and failures so that you can read the chapter to meet your learning needs.

1. The major problem leading to heat-related death is
   a. overdressing for the temperature.
   b. lack of fans.
   c. inadequate water intake.
   d. social isolation.
2. Fluid and electrolyte balance implies that the quantities of these substances entering the body ______________ the quantities leaving the body.
3. Which of the following statements about fluid and electrolyte balance is/are true?
   a. Fluid balance is independent of electrolyte balance.
   b. The concentration of an individual electrolyte is the same throughout the body.
   c. Water and electrolytes occur in compartments in which the composition of fluid varies.
   d. Water is evenly distributed throughout the tissues of the body.

4. About 63% of the total body fluid occurs within the cells in a compartment called the _______________ fluid compartment.

5. Blood and cerebrospinal fluid occur in the _______________ fluid compartment.

6. Which of the following electrolytes is/are most concentrated within the cell?
   a. sodium
   b. bicarbonate
   c. chloride
   d. potassium

7. Plasma leaves the capillary at the arteriole end and enters interstitial spaces because of _______________ pressure.

8. Fluid returns from the interstitial spaces to the plasma at the venule ends of the capillaries because of _______________ pressure.

9. The most important source of water for the normal adult is
   a. in the form of beverages.
   b. from moist foods such as lettuce and tomatoes.
   c. from oxidative metabolism of nutrients.

10. Thirst is experienced when
    a. the mucosa of the mouth begins to lose water.
    b. salt concentration in the cell increases.
    c. the hypothalamus is stimulated by the increasing osmotic pressure of extracellular fluid.
    d. the cortex of the brain is stimulated by shifts in the concentration of sodium.

11. The primary regulator of water output is through
    a. loss in the feces.
    b. evaporation as sweat.
    c. urine production.
    d. loss via respiration.

12. When the renal tubules fail to reabsorb water, the individual has a condition known as _______________ _______________.

13. As dehydration develops, water is lost first from the _______________ _______________ compartment.

14. In treating dehydration, it is necessary to replace
    a. amino acids.
    b. glucose.
    c. water.
    d. electrolytes.

15. In water intoxication, water (enters/leaves) the cell.

16. Which of the following are at increased risk of dehydration because of the inability of the kidneys to concentrate urine?
    a. neonates
    b. school-age children
    c. young adults
    d. elderly people

17. Edema is a likely development in people suffering from starvation because lack of adequate protein leads to (increased/decreased) plasma (osmotic/hydrostatic) pressure.

18. The primary sources of electrolytes are _______________ and _______________.

19. List three routes by which electrolytes are lost.

20. Sodium and potassium ion concentration are regulated by the kidneys and the hormone _______________.

21. Parathyroid hormone regulates calcium ion concentration in the plasma by
    a. freeing calcium from bones.
    b. influencing the renal tubule to conserve calcium.
    c. stimulating the absorption of calcium from the intestine.
    d. freeing calcium from muscle tissue.

22. Prolonged diarrhea is likely to result in
    a. low sodium (hyponatremia).
    b. high sodium (hypernatremia).
    c. low potassium (hypokalemia).
    d. high potassium (hyperkalemia).

23. Acid-base balance is mainly concerned with regulating ______________ ______________ concentration.

24. A pH of 8.5 is said to be
   a. acid.
   b. neutral.
   c. alkaline.

25. The pH of arterial blood is normally ______________.

26. Normal metabolic reactions produce (more/less) acid than base.

27. Anaerobic respiration of glucose produces
   a. carbonic acid.
   b. lactic acid.
   c. acetoacetic acid.
   d. ketones.

28. The strength of an acid depends on the
   a. number of hydrogen ions in each molecule.
   b. nature of the inorganic salt.
   c. degree to which molecules ionize in water.
   d. concentration of acid molecules.

29. A base is a substance that will ______________ ______________ hydrogen ions.

30. A buffer is a substance that
   a. returns an acid solution to neutral.
   b. converts acid solutions to alkaline solutions.
   c. converts strong acids or bases to weak acids or bases.
   d. returns an alkaline solution to neutral.

31. The most important buffer system in plasma and intracellular fluid is the ______________ buffer.
   a. bicarbonate
   b. phosphate
   c. protein

32. The respiratory center controls hydrogen ion concentration by controlling the ______________ and ______________ of respiration.

33. The slowest acting of the mechanisms that control pH is the
   a. buffer system.
   b. respiratory system.
   c. kidney.

34. Hydrogen ions are excreted in the urine as ______________ ions.

35. The accumulation of dissolved carbon dioxide is known as
   a. respiratory acidosis.
   b. respiratory alkalosis.
   c. metabolic acidosis.
   d. metabolic alkalosis.

36. Light-headedness, agitation, dizziness, and tingling sensations are symptoms of
   a. respiratory acidosis.
   b. respiratory alkalosis.
   c. metabolic acidosis.
   d. metabolic alkalosis.

STUDY ACTIVITIES

I. Definition of Key Terms

Define the following key terms used in this chapter.

acid

acidosis

alkalosis

base

buffer system

electrolyte balance

extracellular

intracellular

osmoreceptor

transcellular

water balance

II. Introduction and Distribution of Body Fluids (pp. 814-16)

A. Describe the pathophysiologic mechanisms of heat-related death.

B. Because electrolytes are dissolved in water, water and electrolyte balance are _______________.

C. Answer these questions concerning the composition of body fluids.

1. What is the composition of extracellular fluid?

2. What is the composition of intracellular fluid?

D. What mechanisms are responsible for the movement of fluid and electrolytes from one compartment to another? Describe these mechanisms as fully as possible.

III. Water Balance (pp. 816-18)

A. Answer these questions concerning water intake.

1. Does water balance exist when the gain of water from all sources is equal to the total loss of water?

2. What is the usual water intake for an adult in a moderate climate?

3. From what sources is this water derived?

B. Describe the mechanism that regulates the intake of water.

C. What structure allows water to pass through the lipoprotein cell membrane?

D. Answer the following questions about regulation of water loss.

1. By what routes is water lost from the body?
2. Explain the action of ADH (antidiuretic hormone).
3. What is a diuretic?
4. How do diuretics produce their effects?

E. Answer these questions concerning dehydration.

1. Define dehydration.
2. What conditions may lead to dehydration?
3. How does age influence the development of dehydration?
4. What is the treatment for dehydration?

F. Answer these questions concerning water intoxication.

1. What is water intoxication?
2. What are the symptoms of water intoxication?
3. How is water intoxication treated?

H. Answer these questions concerning edema.

1. What is edema?
2. List four causes of edema.

IV. Electrolyte Balance (pp. 819-23)

A. Answer these questions concerning electrolyte intake.

1. List the electrolytes that are important to cellular function.

2. What are the sources of these electrolytes?

B. Ordinarily, sufficient amounts of fluid and electrolytes are obtained as an individual responds to _______________ and _______________.

C. List three routes by which electrolytes are lost.

D. Answer these questions concerning the regulation of electrolyte balance.

1. What is the role of aldosterone in electrolyte regulation?

2. What is the role of parathyroid hormone in electrolyte regulation?

3. What is the role of the kidney in electrolyte regulation?

4. How is the concentration of negatively charged ions regulated?

E. Describe the causes and symptoms of the following electrolyte problems.

1. high calcium concentration (hypercalcemia)

2. low calcium concentration (hypocalcemia)

3. low sodium concentration (hyponatremia)

4. high sodium concentration (hypernatremia)

5. low potassium concentration (hypokalemia)

6. high potassium concentration (hyperkalemia)

V. Acid-Base Balance (pp. 823-28)

A. Answer these questions concerning acid-base balance.

1. What is an acid?

2. What is a base?

3. What is acid-base balance?

4. How is the concentration of hydrogen ions expressed?

B. List and describe the sources of hydrogen ions in the body.

C. What is the difference between a strong acid or base and a weak acid or base?

D. Answer these questions concerning the regulation of hydrogen ion concentration.

1. How is hydrogen ion concentration regulated by acid-base buffer systems?

2. Describe how this works with the bicarbonate buffer system, the phosphate buffer system, and the protein buffer system.

3. How does the respiratory center regulate hydrogen ion concentration?

4. How do the kidneys help regulate hydrogen ion concentration?

5. How do these mechanisms differ from each other, especially with respect to speed of action?

VI. Acid-Base Imbalances (pp. 829-31)

A. What is the normal arterial blood pH?

B. Fill in the following chart.

Acid-base disturbances			
Disturbance	**Etiologic factors**	**Compensatory mechanisms**	**Symptoms**
Respiratory acidosis			
Metabolic acidosis			
Respiratory alkalosis			
Metabolic alkalosis			

C. What is meant by compensated and uncompensated acidosis or alkalosis?

VII. Clinical Focus Question

Your family is planning a day excursion to the seashore. The age range of your group is from your sister's two-month-old' daughter to your 75-year-old grandparents. The temperature is expected to be in the 90s. In helping your mother plan the food and beverages to bring, what suggestions would you make to maintain everyone's fluid and electrolyte balances?

When you have completed the study activities to your satisfaction, retake the mastery test and compare your performance with your initial attempt. If there are still areas you do not understand, repeat the appropriate study activities.

CHAPTER 22
REPRODUCTIVE SYSTEMS

OVERVIEW

This chapter explains the reproductive system—a unique system because it is essential for the survival of the species rather than for the survival of the individual. The structures and functions of the male and female reproductive systems are explained (objectives 1, 2, 7, and 8). This chapter describes the structures of the ovary and testis, and how they produce sex cells (objectives 3 and 9). It traces the paths of both the egg and sperm, and describes the hormonal controls and accessory organs needed for the sperm and egg to unite (objectives 4–6 and 10). The chapter describes the events of the menstrual cycle (objective 11), and it explains the major events of pregnancy (objectives 12–16).

Knowledge of the anatomy and physiology of the male and female reproductive systems is basic to the study of human sexuality and the process of reproduction.

CHAPTER OBJECTIVES

After you have studied this chapter, you should be able to:

1. State the general functions of the male reproductive system.
2. Name the parts of the male reproductive system and describe the general functions of each part.
3. Outline the process of spermatogenesis.
4. Trace the path sperm cells follow from their site of formation to the outside.
5. Describe the structure of the penis, and explain how its parts produce an erection.
6. Explain how hormones control the activities of the male reproductive organs and the development of male secondary sex characteristics.
7. State the general functions of the female reproductive system.
8. Name the parts of the female reproductive system and describe the general functions of each part.
9. Outline the process of oogenesis.
10. Describe how hormones control the activities of the female reproductive organs and the development of female secondary sex characteristics.
11. Describe the major events that occur during a menstrual cycle.
12. Describe the hormonal changes in the maternal body during pregnancy.
13. Describe the birth process, and explain the role of hormones in this process.
14. List several methods of birth control and describe the relative effectiveness of each method.

FOCUS QUESTION

How do the male and female reproductive systems differ and complement each other?

MASTERY TEST

Now take the mastery test. Do not guess. Some questions may have more than one correct answer. As soon as you complete the test, correct it. Note your successes and failures so that you can read the chapter to meet your learning needs.

Questions 1–4. Match the structures listed in the first column with the functions listed in the second column.

Structure	Function
____ 1. primary sex organs (ovaries and testes)	a. produce such changes as growth of body hair, thickened skin, enlarged larynx, thickened and strengthened bone
____ 2. sex cells	b. join to form a new individual
____ 3. estrogen	c. produce such changes as increased deposition of adipose tissue in subcutaneous layer, increased vascularization, growth of pubic and axillary hair
____ 4. testosterone	d. produce hormones and sex cells

5. The primary organ(s) of the male reproductive system is/are the ______________.

6. During fetal life, the testes are located
   a. in the scrotum.
   b. posterior to the parietal peritoneum.
   c. within the abdominal cavity.
   d. in the pelvis.

7. The descent of the testes into the scrotum is stimulated by
   a. increasing pressure within the abdominal cavity.
   b. the male hormone testosterone.
   c. shortening of the gubernaculum.
   d. shifts in core temperature of the fetus.

8. The testes are suspended in the scrotum by the ______________ ______________.

9. The complex network of channels derived from the seminiferous tubules is the
   a. rete testis.
   b. tunica albuginea.
   c. mediastinum testis.
   d. epididymis.

10. Prior to adolescence, the undifferentiated spermatogenic cells in the testes are called ______________.

11. Each primary spermatocyte will eventually produce how many sperm cells? ______________

12. The nucleus in the head of the sperm contains ______________ chromosomes.

13. Spermatogenesis occurs (continuously/episodically) in men.

14. The function of the epididymis is to
    a. produce sex hormones.
    b. provide the sperm with mobile tails.
    c. store sperm as they mature.
    d. supply some of the force needed for ejaculation.

15. The vas deferens passes through the ______________ ______________ just before emptying into the urethra.

16. Which of the following substances are added to sperm cells by the seminal vesicle?
    a. acid
    b. fructose
    c. glucose
    d. prostaglandins

17. The secretion of the prostate gland is added to seminal fluid during ______________.

18. Which of the following factors is *not* a risk factor for enlargement of the prostate gland in men?
    a. a diet high in fat
    b. frequent use of condoms
    c. exposure to cadmium
    d. having had a vasectomy

19. The function of the bulbourethral glands is to
    a. neutralize the acid secretions of the vagina.
    b. nourish sperm cells.
    c. lubricate the penis.
    d. increase the volume of semen.

20. The external organs of the male reproductive system include the
   a. penis.
   b. testes.
   c. prostate gland.
   d. scrotum.
21. The smooth muscle in the scrotum is the _______________ muscle.
22. Erection of the penis depends on
   a. contraction of the perineal muscles.
   b. filling of the corpus spongiosum with arterial blood.
   c. enlargement of the glans penis.
   d. peristaltic contractions of the vas deferens.
23. Hormones that control male reproductive functions are secreted by the _______________, the _______________, and the _______________ _______________.
24. The pituitary hormone that stimulates the testes to produce testosterone is
   a. gonadotropin-releasing hormone.
   b. FSH.
   c. LH (ICSH).
   d. ACTH.
25. In the male, the growth of body hair, especially in the axilla, face, and pubis, and increased muscle and bone development are examples of _______________ _______________ characteristics.
26. Oversecretion of FSH is prevented by secretion of the hormone _______________.
27. The primary organs of the female reproductive system are the _______________.
28. In the ovary, the primary germinal epithelium is located
   a. in the medulla.
   b. between the medulla and cortex.
   c. in the cortex.
   d. on the free surface of the ovary.
29. The primary oocyte produces _______________ mature egg cell(s).
30. At puberty, the primary oocyte matures within the _______________ _______________.
31. The formation of polar bodies during oogenesis is a rare example of wasted energy and cellular material.
   a. True
   b. False
32. The egg cell is nourished by the
   a. theca interna.
   b. theca externa.
   c. zona pellucida.
   d. corona radiata.
33. The egg is released by the ovary in a process called _______________.
34. Which of the following statements is/are true about the fallopian (uterine) tubes?
   a. The end of the uterine tube near the ovary has many fingerlike projections called fimbriae.
   b. The fimbriae are attached to the ovaries.
   c. The inner layer of the ovarian tube is cuboidal epithelium.
   d. There are cilia in the lining of the fallopian tube that help move the egg toward the uterus.
35. The inner layer of the uterus is the _______________.
36. The upper portion of the vagina that surrounds the cervix is the
   a. fornix.
   b. rectouterine pouch.
   c. vestibule.
   d. hymen.
37. Which of the following statements about the vagina is/are *true?*
   a. The mucosal layer contains many mucous glands.
   b. The levator ani is primarily responsible for closing the vaginal orifice.
   c. The mucosa is drawn into many longitudinal and transverse ridges.
   d. The hymen is a membrane that covers the mouth of the cervix.
38. The organ of the female reproductive system that is analogous to the penis is the
   a. vagina.
   b. mons pubis.
   c. clitoris.
   d. labia majora.

39. Which of the following tissues become engorged and erect in response to sexual stimulation?

a. clitoris
b. labia minora
c. outer third of the vagina.
d. upper third of the vagina.

40. The hormonal mechanisms that control female reproductive functions are (more/less) complex than those of the male.

41. The primary female sex hormones are _______________ and _______________.

42. Which of the following secondary sex characteristics in the female seem to be related to androgen concentration?

a. breast development
b. growth of axillary and pubic hair
c. female skeletal configuration
d. deposition of adipose tissue over hips, thighs, buttocks, and breasts

43. Which of the following is *not* a source of female sex hormones?

a. pituitary gland
b. placenta
c. ovary
d. adrenal gland

44. During the menstrual cycle, the event that seems to provoke ovulation is

a. increasing levels of progesterone.
b. a sudden increase in concentration of LH.
c. decreasing levels of estrogen.
d. a cessation of secretion of FSH.

45. After the release of an egg, the follicle forms a _______________ _______________.

46. As the above structure develops, the level of which of the following hormones increases?

a. estrogen
b. progesterone
c. FSH
d. LH

47. As the hormone levels change in the part of the cycle before and immediately after ovulation, which of the following changes are seen in the uterus?

a. growth of the myometrium
b. increase in adipose cells of the perimetrium
c. thickening of the endometrium
d. decrease in uterine gland activity

48. The concentration of which of the following hormones decreases following ovulation?

a. estrogen
b. progesterone
c. FSH
d. LH

49. The beginning of the menstrual cycle at puberty is called _______________; the cessation of the menstrual cycle in middle age is called _______________.

50. Fertilization of the egg by the sperm takes place in the

a. vagina.
b. cervix.
c. uterus.
d. fallopian tube.

51. Which of the following is thought to be the mechanism by which the sperm enters the egg?

a. An antigen-antibody reaction that briefly alters the cell membrane of the egg.
b. The structure of the cell membrane of the egg allows entry.
c. The head of the sperm has an enzyme that permits digestion through the corona radiata and zona pellucida.
d. The mechanism is unknown.

52. Implantation takes place by the end of _______________ week(s) following fertilization.

53. The hormone that maintains the corpus luteum following implantation is

a. LH.
b. HCG.
c. progesterone.
d. FSH.

54. The primary source of hormones needed to support a pregnancy after the first three months is the _______________.

55. Which of the following will *not* increase during pregnancy?

a. blood volume
b. hematocrit
c. cardiac output
d. urine production

56. Labor is initiated by a decrease in _______________ levels and the secretion of _______________ by the posterior pituitary gland.

57. Bleeding following expulsion of the afterbirth is controlled by

a. hormonal mechanisms.
b. increased fibrinogen levels.
c. contraction of the uterine muscles.
d. sympathetic stimulation of arterioles.

58. Milk production is stimulated by the hormone _______________.

59. The most common breast disease in women is

a. fibrocystic disease.
b. breast cancer.
c. mastitis.
d. silicosis secondary to breasts implants.

60. A drug that is currently under investigation for its ability to block the development of breast cancer in women with a strong family history of the disease is _______________.

61. Milk is secreted from the breast

a. as production fills the duct structure.
b. in response to hormonal stimulation.
c. in response to mechanical stimulation of the nipple; i.e., sucking.
d. in response to gonadotropins.

62. It is possible to become pregnant while breastfeeding.

a. True
b. False

63. Methods of contraception that utilize mechanical barriers are

a. coitus interruptus.
b. the condom.
c. the diaphragm.
d. the IUD.

64. The method of contraception that is thought to interfere with ovulation is

a. the rhythm method.
b. the oral contraceptive (the pill).
c. chemical foams.
d. the IUD.

65. The commonly used surgical technique to sterilize the male is the _______________.

66. The contraceptive method that is also effective in preventing the spread of sexually transmitted diseases (STDs) is the

a. pill.
b. coitus interruptus.
c. condom.
d. IUD.

67. Sexually transmitted diseases are not treated promptly primarily because

a. patients are embarrassed to consult a physician.
b. there is a perception that they can be effectively treated with home remedies.
c. many of the symptoms of STDs are similar to symptoms of diseases or allergies that are not sexually related.
d. of ignorance of how they are transmitted.

Study Activities

I. Definition of Key Terms

Define the following key terms used in this chapter.

androgen

cleavage

clitoris

coitus

contraception

ejaculation

emission

epididymis

estrogen

fertilization

fimbriae

follicle

gonadotropin

gubernaculum

implantation

infundibulum

inguinal

meiosis

menopause

menstrual cycle

oogenesis

orgasm

ovulation

placenta

pregnancy

progesterone

puberty

semen

spermatogenesis

testosterone

vas deferens

zygote

II. Introduction and Organs of the Male Reproductive System (p. 836)

A. Describe the functions of the reproductive system.

B. Answer the following concerning the organs of the male reproductive system.

1. Label the structures in the accompanying diagram: urinary bladder, prostate gland, penis, urethra, glans penis, testis, vas deferens, seminal vesicle, ampulla, ejaculatory duct, bulbourethral gland, epididymis, ureter.

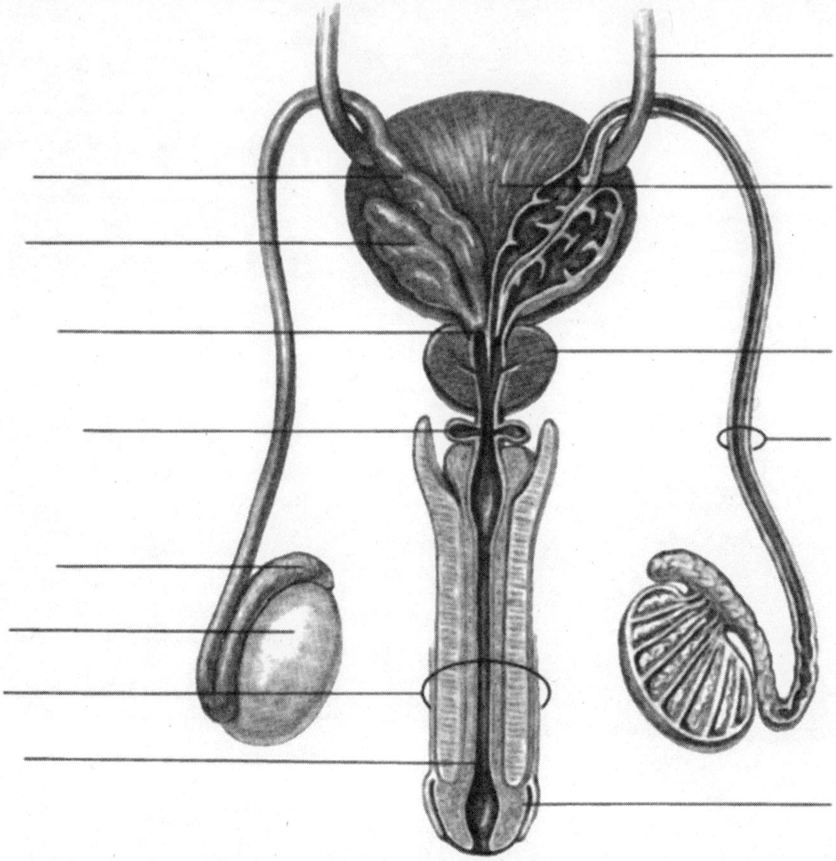

2. What is the primary function of the male reproductive system?

3. What are the primary organs of the male reproductive system?

III. Testes (pp. 836-42)

A. Answer the following questions concerning the descent of the testes.

1. Where do the testes originate during fetal life?

2. How do the testes descend into the scrotum?

3. What is cryptorchidism? How does this produce sterility?

4. How does an inguinal hernia develop?

B. Answer these questions concerning the structure of the testes.

1. What is the function of spermatogenic cells?

2. What is the function of interstitial cells?

3. How is a testicular cancer found on physical examination? How is it treated?

C. Describe the process of sperm formation. Include the events of puberty, where sperm are stored, and a description of the sperm.

D. Answer these questions concerning the effects of toxic substances on sperm.

1. What is the usual effect of toxic substances on sperm?

2. What chemical is an exception?

IV. Male Internal Accessory Organs (pp. 842-44)

A. List the internal accessory organs of the male reproductive system.

B. Answer the following concerning the epididymis.

1. Describe the location and structure of the epididymis.

2. What is the function of the epididymis in sperm formation?

3. What is the function of the epididymis in the emission of sperm?

C. Describe the course of the vas deferens.

D. Answer these questions concerning the seminal vesicles.

1. Where are the seminal vesicles located?

2. What is the nature and the function of the secretion of the seminal vesicles?

E. Answer these questions concerning the prostate gland.

1. Where is the prostate gland located?

2. What is the nature and the function of the secretion of the prostate gland?

3. When is this secretion released?

4. What factors increase the risk of prostatic enlargement?

5. What diagnostic tests are used to diagnose prostatic enlargement?

F. Answer the following concerning the bulbourethral glands.

1. Describe the location and function of the bulbourethral glands (Cowper's glands).

2. What is the stimulus for the release of this secretion?

G. Describe semen.

H. List the possible causative factors for male infertility and describe the diagnostic procedures used to diagnose male infertility.

V. Male External Reproductive Organs (pp. 845-48)

A. Answer the following questions about the scrotum.

1. Of what tissues is the scrotum composed?

2. What is the result of the contraction of the dartos muscle?

B. Label the following structures on the drawing of a cross section of a penis: dorsal vein, dorsal nerve, dorsal artery, deep artery, corpora cavernosa, tunica albuginea, urethra, corpus spongiosum, connective tissue, subcutaneous tissue, skin.

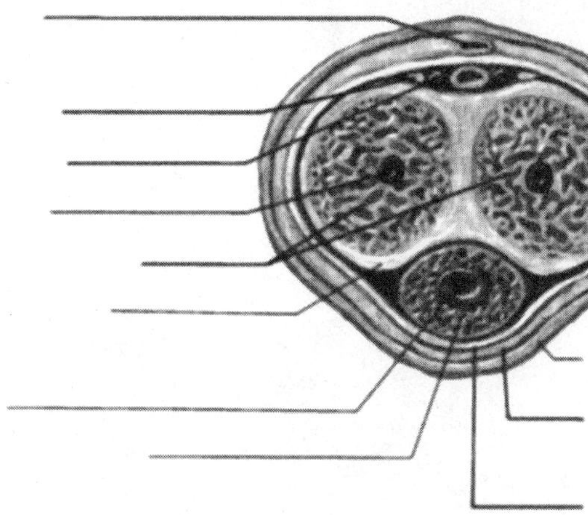

C. Removal of the prepuce is called _______________.

D. Describe the events of erection and ejaculation. Be sure to distinguish between emission and ejaculation.

VI. Hormonal Control of Male Reproductive Functions (pp. 848-50)

A. Answer these questions concerning pituitary hormones.

1. What glands secrete hormones that control male reproductive functions?

2. What is the function of FSH and LH?

B. Answer the following concerning male sex hormones.

1. Where is testosterone produced?

2. What is the function of testosterone?

3. List the male secondary sexual characteristics.

C. Describe the regulation of sex hormones in the male.

VII. Organs of the Female Reproductive System (p. 850)

Label the structures in the accompanying drawing: uterine tube, fimbriae, ovary, uterus, urinary bladder, symphysis pubis, urethra, clitoris, labium min, labium maj, vaginal orifice, fornix, cervix, rectum, vagina, anus.

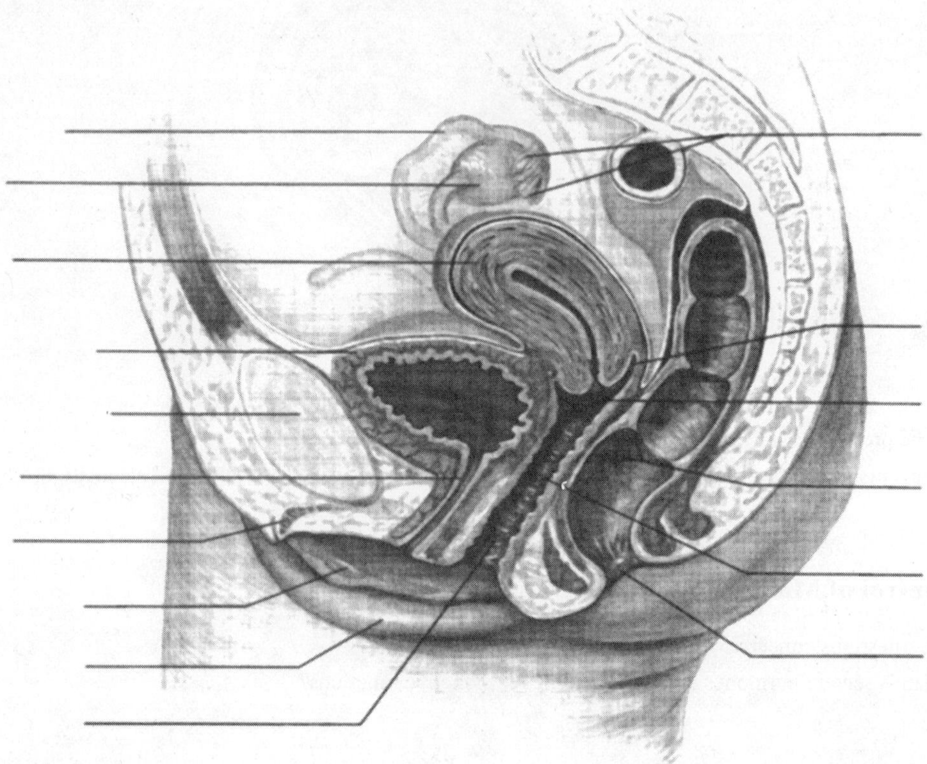

VIII. Ovaries (pp. 850-55)

A. Answer the following concerning the descent of the ovaries.

1. Where do the ovaries originate during fetal life?

2. Describe the descent of the ovaries to their position in the pelvic cavity.

B. Describe the structure of the ovary.

C. Answer the following concerning the formation of egg cells.

1. Outline the process of egg cell production.

2. How is this different from spermatogenesis?

D. Answer the following concerning the maturation of a follicle.

1. What stimulates the maturation of a primary follicle at puberty?

2. What changes occur in the follicle as a result of maturation?

3. Describe the structure of the egg within the follicle.

4. Why is increased maternal age associated with increased incidence of chromosomal defects in the children of older women?

5. Describe the procedure known as polar body biopsy and how the results are interpreted.

E. Answer these questions concerning ovulation.

1. What provokes ovulation?

2. What happens to the egg after it leaves the ovary?

IX. Female Internal Accessory Organs (pp. 855-58)

A. Label the structures in the accompanying drawing: body of uterus, uterine tube, infundibulum, cervical orifice, ovary, broad ligament, round ligament, ovarian ligament, cervix, vagina, fimbriae, secondary oocyte, follicle, endometrium, myometrium, perimetrium.

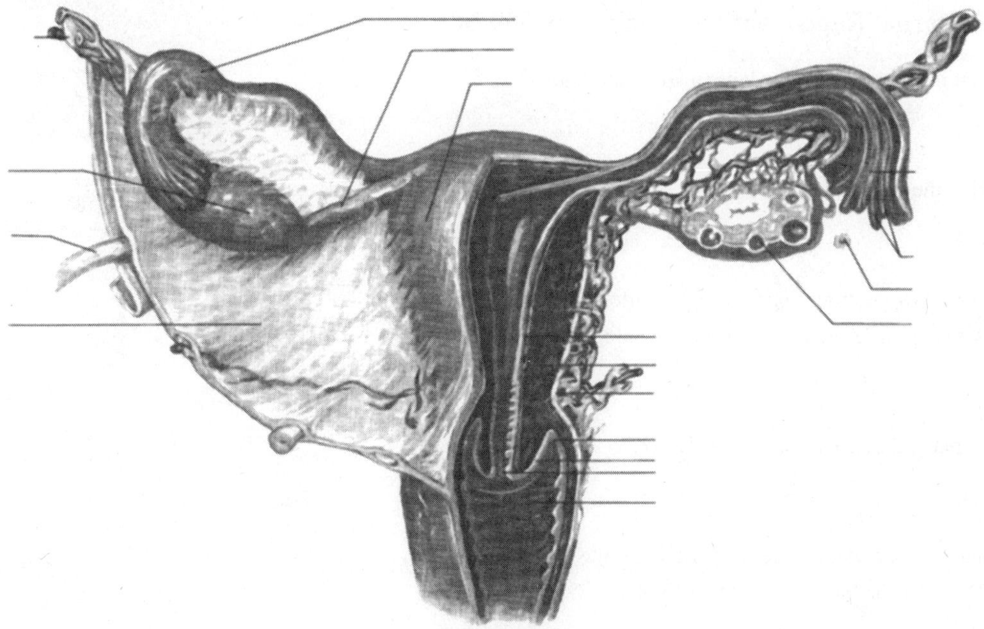

B. How does the structure of the uterine tube move the egg toward the uterus?

C. Answer the following concerning the uterus.

1. Based on the description in the text, draw a uterus. Label the body, cervix, endometrium, myometrium, and perimetrium.

2. What is a Pap smear?

3. How are the muscle fibers arranged in the myometrium?

D. Answer the following concerning the vagina.

1. What is the function of the vagina?

2. What are the fornices and the hymen?

3. Describe the structure of the vagina.

4. What has occurred in many of the daughters of women who took diethylstilbestrol during pregnancy?

X. Female External Reproductive Organs (pp. 859-60)

A. To what male organs are the labia majora analogous?

B. Describe the labia minora.

C. Answer these questions concerning the clitoris.

1. To what male organ is the clitoris analogous?

2. What is the structure of the clitoris?

D. Answer the following concerning the vestibule.

1. Describe the location of the vestibule.

2. What is the function of the vestibular glands (Bartholin's glands)?

3. To what male structures are the vestibular glands and vestibular bulbs analogous?

E. Describe the events of erection and orgasm in the female.

XI. Hormonal Control of Female Reproductive Functions (pp. 860-64)

A. Answer these questions concerning pituitary hormones.

1. What appears to initiate sexual maturation in the female?

2. What pituitary hormones influence sexual function in the female?

3. What are the differences in the hormonal mechanism in the male and the female?

B. Answer these questions concerning female sex hormones.

1. What are the sources of female sex hormones?

2. What is the function of estrogen?

3. What is the function of progesterone?

4. Female athletes may experience a disturbance of the menstrual cycle. This seems to be due to a loss of _______________ _______________ and a decline in _______________.

C. Answer the following concerning female reproductive cycles.

1. What is menarche?

2. Describe the events of the menstrual cycle. Include shifts in hormone levels, uterine changes, and ovarian changes.

D. Answer these questions concerning menopause.

1. What is menopause?

2. What seems to be the cause of menopause?

E. How can a woman bear a child after menopause?

XII. **Pregnancy** (pp. 864-71)

A. Answer these questions concerning the transport of sex cells.

1. How are egg cells transported in the uterine tubes?

2. How are sperm transported through the uterus and along the uterine tubes?

3. How long do egg and sperm cells survive?

B. Answer these questions concerning fertilization.

1. How does the sperm fertilize the egg?

2. What is "zona blasting"?

3. What is the name of the cell that is formed during fertilization?

C. Describe the developmental events from fertilization to implantation.

D. Answer these questions concerning hormonal changes during pregnancy.

1. What mechanism maintains the corpus luteum during early pregnancy?

2. What hormones are secreted by the placenta?

3. What are the roles of relaxin, estrogen, and progesterone during pregnancy?

E. Describe the physical changes experienced by the mother as the fetus grows.

F. How do pregnancy test kits work?

G. Answer these questions concerning the birth process.

1. What changes take place in placental secretion after the seventh month of gestation?

2. What stimulates the secretion of oxytocin?

3. What regulates the secretion of oxytocin?

4. What is labor?

5. What is afterbirth?

6. How is bleeding controlled once the uterus is emptied?

7. What is involution?

H. Describe the causes of female infertility.

XIII. **Mammary Glands** (pp. 871-77)

A. What is the function of the mammary glands?

B. Where are the mammary glands located?

C. Describe the structure of the mammary glands.

D. Discuss the detection, treatment, and prevention of breast cancer.

E. Answer these questions concerning the development of breasts.

1. How does puberty affect the breast in the male and in the female?

2. How does pregnancy affect the breast?

F. Describe the production and secretion of milk.

G. What are the advantages of breast-feeding?

XIV. **Birth Control** (pp. 877-81)

Describe each of the following methods of birth control, and identify the advantages and disadvantages of each method.

coitus interruptus

rhythm method

condom

diaphragm

cervical cap

vaginal sponge

chemical barriers

oral contraceptives

"morning after" pill

contraceptive implant

intrauterine devices

vasectomy

tubal ligation

XV. **Sexually Transmitted Diseases** (pp. 882-83)

List the common symptoms of sexually transmitted diseases.

XVI. Clinical Focus Question

Your local school board is embroiled in a controversy related to when and if sex education should be offered in the school system and, if offered, when should it begin and what should be the goals of such education. Based on your knowledge of the male and female reproductive systems and on your own value system, develop a personal position on these questions.

When you have completed the study activities to your satisfaction, retake the mastery test and compare your performance with your initial attempt. If there are still areas you do not understand, repeat the appropriate study activities.

CHAPTER 23
HUMAN GROWTH AND DEVELOPMENT

OVERVIEW

This chapter is about growth, which is an increase in size, and development, a process by which an individual changes from one life phase to another (objective 1). This chapter explains the major events of prenatal life: the period of cleavage, the origin of primary germ layers, the formation and function of the placenta, and the events of the fetal stage (objectives 2–6). The physiological shifts necessary for a newborn while adjusting to postnatal life are discussed (objective 7), and the stages of development from birth to death are described (objective 8).

Knowledge of growth and development complements the knowledge of the reproductive system and the understanding of how the species survives.

CHAPTER OBJECTIVES

After you have studied this chapter, you should be able to:

1. Distinguish between growth and development.
2. Describe the major events of the period of cleavage.
3. Explain how the primary germ layers originate and list the structures each layer produces.
4. Describe the formation and function of the placenta.
5. Define *fetus* and describe the major events that occur during the fetal stage of development.
6. Trace the general path of blood through the fetal circulatory system.
7. Describe the major circulatory and physiological adjustments that occur in the newborn.
8. Name the stages of development between the neonatal period and death, and list the general characteristics of each stage.

FOCUS QUESTION

Are patterns of growth and development consistent enough from one individual to another to allow for prediction of behavior?

MASTERY TEST

Now take the mastery test. Do not guess. Some questions may have more than one correct answer. As soon as you complete the test, correct it. Note your successes and failures so that you can read the chapter to meet your learning needs.

1. An increase in size is called _______________.
2. The process by which an individual changes throughout life is _______________.
3. The period of life that begins with fertilization and ends at birth is known as the _______________ period.
4. The period of life from birth to death is known as the _______________ period.
5. After fertilization, the zygote divides into smaller and smaller cells, forming a solid ball called the
   a. blastomere.
   b. blastocyst.
   c. morula.
   d. embryo.
6. The above structure is enclosed in the
   a. zona pellucida.
   b. placenta.
   c. chorion.
   d. amnion.
7. The morula can be the source of a cell used to diagnose genetic disease.
   a. True
   b. False

8. Which of the following statements is/are true about implantation?

a. Implantation is complete by one month after fertilization.

b. Implantation marks the end of the period of cleavage.

c. The fertilized egg adheres to the wall of the uterus, and the endometrium then grows over it.

d. The trophoblast produces fingerlike projections that penetrate the endometrium.

9. A woman who has lost both fallopian tubes can become pregnant using a technique called _______________ _______________ fertilization.

10. The ectoderm, mesoderm, and endoderm are _____________ _____________layers.

11. From which of the layers of the embryonic disk do the hair, nails, and glands of the skin arise?

a. endoderm

b. ectoderm

c. mesoderm

12. Which of the following structures arises from the mesoderm?

a. lining of the mouth

b. muscle

c. lining of the respiratory tract

d. epidermis

13. At what time does the embryonic disk become a cylinder?

a. four weeks of development

b. six weeks of development

c. eight weeks of development

d. four days of development

14. The chorion in contact with the endometrium becomes the _______________.

15. The membrane covering the embryo is called the _______________.

16. The number of blood vessels in the umbilical cord is

a. one artery and one vein.

b. one artery and two veins.

c. two arteries and one vein.

d. two arteries and two veins.

17. The embryonic structure that forms fetal blood cells is the

a. amnion.

b. placenta.

c. allantois.

d. yolk sac.

18. The embryonic stage ends at _______________ weeks.

19. Factors that cause congenital malformations by affecting the embryo are called _______________.

20. A defect that can occur on day 28 in development is a _______________ _______________.

21. Fetal skeletal muscles are active enough to permit the mother to feel fetal movements in month

a. four.

b. five.

c. six.

d. seven.

22. Brain cells of the fetus rapidly form networks during the

a. first trimester.

b. second trimester.

c. third trimester.

d. all three trimesters.

23. Which of the following factors indicate's that the respiratory system is mature enough to allow survival when a baby is born prematurely?

a. gestational age of seven months

b. thinness of the respiratory membrane

c. respiratory rate below 40 per minute

d. sufficient amounts of surfactant

24. Oxygen and nutrient-rich blood reach the fetus from the placenta via the umbilical _______________.

25. The ductus venosus shunts blood around the

a. liver.

b. spleen.

c. pancreas.

d. small intestine.

26. The structures that allow blood to avoid the nonfunctioning fetal lungs are the _______________ _______________ and the _______________ _______________.

27. The factor that decreases the effort required for an infant to breathe after the first breath is _______________.

28. The primary energy source for the newborn is
   a. glucose.
   b. fat.
   c. protein.
29. An infant who is breast-fed receives milk
   a. immediately after birth.
   b. within 12 hours after birth.
   c. within 48 hours after birth.
   d. within 2–3 days after birth.
30. An infant's urine is (more/less) concentrated than an adult's.
31. Which of the following fetal structures closes as a result of a change in pressure?
   a. ductus venosus
   b. ductus arteriosus
   c. umbilical vessels
   d. foramen ovale
32. The beginning of the ability to communicate is accomplished during ______________.
33. Reproductive maturity occurs during ______________.
34. The process of growing old is called ______________.
35. Passive aging is a process of
   a. degeneration.
   b. programmed cell death.
   c. genetic breakdown.
   d. all of the above.
36. Life span and life expectancy are synonomous terms.
   a. True
   b. False

Study Activities

I. Definition of Key Terms

Define the following key terms used in this chapter.

allantois

amnion

apaptosis

chorion

cleavage

embryo

fetus

germ layer

neonatal

placenta

postnatal

prenatal

senescence

umbilical cord

zygote

II. Introduction and Prenatal Period (pp. 892-913)

A. Answer the following concerning the prenatal period.

1. Define *growth.*

2. Define *development.*

3. Describe the prenatal period.

4. Describe the postnatal period.

B. Describe the period of cleavage.

C. How do dizygotic and monozygotic twins occur?

D. Answer these questions concerning estimates of time of conception and birth.

1. Calculate the date of conception when the first day of the last menstrual period is 6/13/95.

2. Calculate the expected date and the range of possible dates of birth for the above example.

3. How is ultrasound used to calculate dates of conception and expected birth?

E. Describe a method to diagnose genetic defects before the implantation of the embryo.

F. Answer the following concerning the embryonic stage.

1. What are the boundaries that define the embryonic stage of development?

2. When does implantation occur?

3. What is the embryonic disk?

4. What are the primary germ layers?

5. List the structures that arise from the
ectoderm

mesoderm

endoderm

6. Describe the events of the fourth through the seventh weeks of development.

7. Describe how in vitro fertilization is accomplished. Include a description of some of the legal problems that can develop with this technique.

8. Label the structures in the accompanying illustration: uterine wall, placenta, umbilical cord, umbilical vein, umbilical arteries, embryonic blood vessels, chorion, villi, lacunae, decidua basalis, maternal blood vessels.

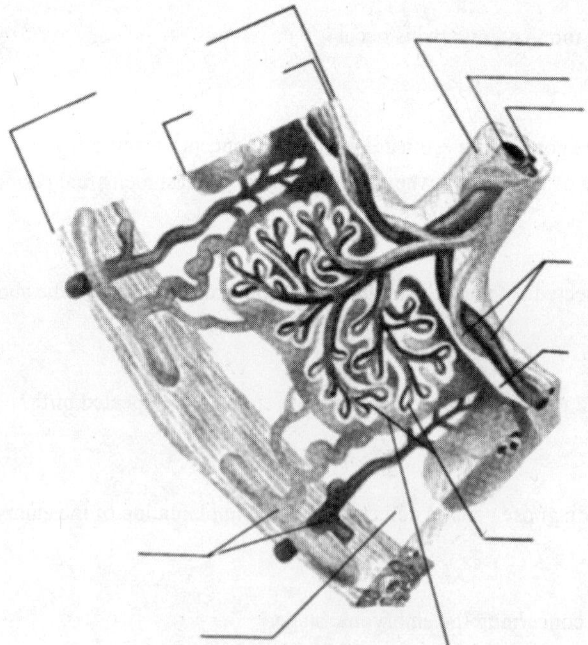

9. Describe the structure and function of the umbilical cord and the fetal membranes.

10. What are the functions of the allantois and the yolk sac?

11. Why is the embryonic stage described as the most critical period of development?

G. Answer the following concerning the fetal stage.

1. What is a teratogen? List some common teratogens.

2. How does a fetus become addicted to drugs?

3. Fill in the following chart.

The fetal stage

Month	Major events of growth and development
Third lunar month	
Fourth lunar month	
Fifth lunar month	
Sixth lunar month	
Seventh lunar month	
Eighth lunar month	
Ninth lunar month	
Tenth lunar month	

4. What kinds of fetal surgery have been performed?

H. Trace a drop of fetal blood from the placenta through the circulatory system. Identify the difference from postnatal circulation.

III. Postnatal Period (pp. 913-22)

A. Fill in the following chart.

The neonatal period	
Life function	**Major changes needed to adjust to extrauterine life**
Respiration	
Nutrition	
Urine formation	
Temperature control	
Circulation	

B. What are the major growth and developmental events of infancy?

C. What are the normal growth and developmental events of childhood?

D. What are the normal growth and developmental events of adolescence?

E. What are the developmental events of adulthood?

F. Describe senescence.

IV. Aging (pp. 922-24)

A. How long is the human life span? How long is human life expectancy?

B. Compare passive and active aging processes.

V. Clinical Focus Question

Your best friend has told you that she is about 2 weeks pregnant. She is quite sure because she has tested her urine using a home pregnancy kit. She tells you that there is little reason for her to seek prenatal care until she is four or five months pregnant. Based on your knowledge of prenatal growth and development, how would you respond?

When you have completed the study activities to your satisfaction, retake the mastery test and compare your performance with your initial attempt. If there are still areas you do not understand, repeat the appropriate study activities.

CHAPTER 24
GENETICS

OVERVIEW

This chapter is about genetics, a field of study that attempts to explain the similarities and differences between parents and their offspring. This chapter explains how gene discoveries are related to the study of anatomy and physiology and to clinical practice in health care professions (objective 1). It also explains how heredity and environment influence the development of individual characteristics (objective 9). It distinguishes between genes and chromosomes and describes the various processes by which parents genetic material is shared by their offspring (objectives 3, 4, 5, 6, 7, and 8). The chapter also describes the effect of gender on inheritance (objectives 10 and 11). Health consequences of deviations in genetic process are described, as well as the challenges presented in diagnosing inherited diseases and the use of gene therapy (objectives 14 and 15).

CHAPTER OBJECTIVES

After you have studied this chapter, you should be able to:

1. Explain how gene discoveries are relevant to the study of anatomy and physiology and to health care.
2. Distinguish between genes and chromosomes.
3. Outline the process of meiosis and explain why meiotic products have different combinations of parental genes.
4. Define the two types of chromosomes.
5. Explain how genes can have many alleles (variants), but a person can have only two alleles of a particular gene.
6. Describe how alleles interact.
7. Distinguish between modes of inheritance by which traits pass from parent to offspring.
8. Explain how gene expression varies among individuals.
9. Describe how multiple genes and the environment interact to produce complex traits.
10. Describe how genes and chromosomes control the inheritance of gender and how traits are transmitted on the X chromosome.
11. Describe how gender affects gene expression.
12. Explain how deviations in chromosome number or arrangement can harm health.
13. Explain how conditions caused by extra or missing chromosomes reflect a meiotic error.
14. Describe the challenges that the ability to diagnose inherited disease presents.
15. Explain how gene therapy works.

FOCUS QUESTION

How is it possible to modify the health outcomes of your genetic makeup?

MASTERY TEST

Now take the mastery test. Do not guess. Some questions may have more than one correct answer. As soon as you complete the test, correct it. Note your successes and failures so that you can read the chapter to meet your learning needs.

1. The source of many new genetic discoveries that affect health care is the _______________ _______________ project.
2. The transmission of genetic information from parents to offspring is _______________.
3. The chemical, physical, and biological factors in the surroundings of individuals that influence their characteristics are known collectively as _______________.

4. The portion of the DNA molecule that contains the information for producing particular types of protein is the
   a. allele.
   b. chromosome.
   c. gene.
   d. zygote.
5. The genes in a human cell are known collectively as the _______________ _______________.
6. The nucleus of each cell in the human body contains _______________ chromosomes from each parent.
7. In the process of meiosis, the number of chromosomes
   a. doubles.
   b. remains the same.
   c. is halved.
8. Chromosome pairs 1 22 are called
   a. karyotypes.
   b. autosomes.
   c. nongender chromosomes.
   d. alleles.
9. A zygote that contains a gene for brown eyes and a gene for blue eyes is said to be
   a. dominant.
   b. recessive.
   c. homozygous.
   d. heterozygous.
10. The combination of genes within a zygote and subsequent daughter cells is said to be the individual s
   a. cell type.
   b. phenotype.
   c. genotype.
   d. genetic endowment.
11. A characteristic that can occur in an individual of either gender and only when both parents have a recessive gene for that characteristic is due to
   a. a mutant gene.
   b. an autosomal, recessive gene.
   c. sex-linked inheritance.
   d. incomplete dominance of the dominant gene.
12. When neither of the genes of a pair is dominant or recessive, the genes are said to be _______________.
13. When a genetic disorder produces different sets of symptoms in family members who have the disorder, the condition is called _______________________.
14. When the same phenotype results from the actions of different genes, the phenomenon is called _______________ _________________.
15. The somatic cells of a female have
   a. two Y chromosomes.
   b. an X and a Y chromosome.
   c. two X chromosomes.
16. If the male parent has the genotype *AA,* and the female parent has the genotype *aa,* the offspring s genotype will be
   a. *AA.*
   b. *aa.*
   c. *Aa.*
17. The health consequences of a disease-causing allele are most likely to be serious when the allele is
   a. a mutant.
   b. sex linked.
   c. autosomal recessive.
   d. autosomal dominant.
18. It is thought that recessive alleles that cause illnesses such as sickle cell disease and Tay-Sachs disease persist in the population because they confer the advantage of increased resistance to infectious disease.
   a. True
   b. False
19. An individual who has two disease-causing alleles but remains healthy is said to have a phenotype that is the expression of two _______________ _______________ alleles.
20. When genetic disorders produce a variety of symptoms across different individuals, the phenomenon is called
   a. variable expression.
   b. incomplete dominance.
   c. pleiotropy.
   d. variable penetrance.
21. Most inherited traits are (monogenic/polygenic).

22. The genotype of a male child is

a. XX.
b. YY.
c. XY.

23. Traits determined by recessive genes located on the X chromosomes are said to be _______________ _______________.

24. Nondisjunction, a common cause of chromosomal abnormalities, occurs during

a. meiosis.
b. mitosis.
c. fertilization.
d. cleavage.

25. Down syndrome is due to a _______________ of chromosome 21.

26. The branch of science that deals with the relationship between inheritance and disease is _______________ _______________.

27. Which of the following statements about genetic counseling is/are true?

a. A counselor can predict whether or not a child will have a genetic disorder.
b. A genetic counselor needs a detailed history called a pedigree chart.
c. The decision to have or not to have a child is solely up to the parents.
d. The counselor should share his or her value system with the parents.

28. Information about the genetic condition of a developing fetus can be obtained by using a technique such as _______________ and _______________ _______________ _______________.

29. Gene therapy that intervenes with the fertilized egg is _______________ gene therapy.

Study Activities

I. Definition of Key Terms

Define the following key terms used in this chapter.

allele

autosome

dominant

gene

genetics

genotype

heredity

heterozygous

homozygous

meiosis

multiple allele

mutation

nondisjunction

phenotype

recessive

sex chromosome

II. Introduction and the Emerging Role of Genetics in Medicine (pp. 929-30)

A. Describe the health applications that our knowledge of genetics could make possible in the near future.

B. Define *heredity* and *environment.*

III. Genes and Chromosomes (pp.930-31)

A. Answer these questions concerning genes.

1. What is a gene?

2. What is the function of a gene?

3. Define *human genome.*

B. Answer these questions concerning chromosomes.

1. What is a chromosome?

2. Where are chromosomes located?

3. How many chromosomes are in each cell?

IV. Chromosomes and Sex Cell Production (pp. 931-40)

A. Answer these questions about chromosomes and sex cell production.

1. What are the products of the process of meiosis?

2. Describe the steps in meiosis.

3. What is crossing over?

4. How does meiosis contribute to genetic variability?

5. How are genetic maps and markers useful in the diagnosis and treatment of disease?

B. Answer these questions concerning chromosomes and genes.

1. What is a karyotype?

2. How are chromosomes duplicated?

3. What is an allele?

4. What is a homozygous genotype?

5. What is a heterozygous genotype?

V. Modes of Inheritance (pp. 940-42)

A. Answer the following concerning modes of inheritance.

1. Compare genotype and phenotype

2. Compare wild-type allele and mutant.

B. Answer the following concerning dominant and recessive inheritance.

1. Describe dominant, recessive, autosomal, and sex-linked modes of inheritance.

2. Using *B* for a hypothetical dominant trait and *b* for the recessive trait, identify the genotypes and phenotypes that are possible.

3. How are the medical consequences different for autosomal recessive conditions and autosomal dominant conditions? Give examples of each.

4. Why do recessive alleles such as those for Tay-Sachs disease persist in a population?

C. Answer the following about incomplete dominance.

1. Define *incomplete dominance.*

2. Describe how this concept is demonstrated in sickle cell disease.

3. Define *codominance.*

4. Describe how this concept is demonstrated in the determination of blood type.

VI. Gene Expression (pp. 942-43)

A. Define *penetrance* and *expressivity.*

B. What is pleiotropy?

VII. Complex Traits (pp. 943-45)

A. Compare monogenic and polygenic traits.

B. Define *multifactorial traits.*

VIII. Matters of Sex (pp. 945-47)

A. Answer the following about gender inheritance.

1. How is the gender of offspring determined?

2. What is the role of the *SRY* gene in sex determination?

B. Answer the following concerning sex chromosomes and sex-linked inheritance.

1. Why are some characteristics linked to sex chromosomes?

2. What traits are known to be sex linked?

3. Why are sex-linked traits more common in males?

IX. Gender Effects on Phenotype (pp. 947-48)

A. Describe genomic imprinting.

B. What is fragile X syndrome?

X. Chromosome Disorders (pp. 948-51)

A. Define *polyploidy* and describe its effects.

B. Define *aneuploidy.*

C. Compare the results of trisomy and monosomy.

D. What is nondisjunction?

XI. Medical Genetics (pp. 951-59)

A. How are genetic tests unlike other diagnostic tests?

B. What is the role of a genetic counselor?

C. Describe gene therapy.

XII. Clinical Focus Question

A classmate tells you that she cannot marry because her mother has Huntington's chorea. How would you respond to her statement?

When you have completed the study activities to your satisfaction, retake the mastery test and compare your performance with your initial attempt. If there are still areas you do not understand, repeat the appropriate study activities.

MASTERY TEST ANSWERS

1 Mastery Test Answers

1. c
2. c
3. Latin and Greek
4. anatomy
5. physiology
6. always
7. movement, responsiveness, growth, reproduction, respiration, digestion, absorption, circulation, assimilation, excretion
8. metabolism
9. a, b, c
10. water
11. energy
living matter
12. energy
13. increases
14. pressure
15. respiration
16. a
17. a
18. d
19. is
20. illness
21. b
22. b
23. d
24. a
25. c
26. cell
tissue
organ
organ system
organism
27. axial
28. appendicular
29. dorsal, ventral
30. diaphragm
31. mediastinum
32. c
33. pleural cavity
34. pericardial
35. abdominopelvic
36. c, d, b, b, d, b, b, c, b, a
37. b, c
38. c
39. b
40. body regions
41. sound waves, magnetic
42. b, d
43. is

2 Mastery Test Answers

1. c
2. d
3. e
4. a
5. b
6. a
7. biochemistry
8. Matter is anything that has weight and takes up space. It occurs in the forms of solids, liquids, and gases.
9. elements or atoms
10. compound
11. b
12. bulk
13. enzymes
14. carbon, hydrogen, oxygen, and nitrogen
15. e
16. 1. c, 2. a, 3. b
17. protons
18. protons, neutrons
19. number, weight
20. protons, neutrons
21. b
22. electrons
23. c
24. half-life
25. a
26. b
27. d
28. a
29. a
30. a, b, c, d
31. the same element
32. different
33. molecular
34. structural
35. synthesis, decomposition
36. reversible
37. catalyst
38. 1. b, 2. c, 3. a
39. 7, less than 7, more than 7
40. alkalosis
41. ether, alcohol; water
42. c
43. a. I, b. O, c. O, d. I, e. O, f. I, g. O
44. carbon, oxygen, hydrogen
45. fatty acids, glycerol
46. amino acids
47. a
48. CT scanning

3 Mastery Test Answers

1. d
2. d
3. c
4. b
5. a
6. b
7. egg (ovum)
8. a, b
9. cytoplasm, nucleus
10. b
11. c
12. a
13. c
14. d
15. b
16. b
17. a
18. a
19. a
20. b
21. b
22. endoplasmic reticulum
23. a, d
24. d

25. glycoproteins
26. powerhouses
27. more
28. d
29. b
30. a, c
31. d
32. cilia, flagella
33. c
34. microfilaments, microtubules
35. nucleolus, chromatin
36. d
37. diffusion
38. facilitated diffusion
39. osmosis
40. a
41. filtration
42. active transport
43. pinocytosis
44. phagocytosis
45. b
46. receptor-mediated endocytosis
47. b
48. mitosis
49. a. c, 2. a, 3. d, 4. b
50. interphase
51. differentiation
52. a
53. a. A, b. B, c. A, d. C, e. A
54. a
55. cancer

4 Mastery Test Answers

1. b
2. anabolic metabolism
3. catabolic metabolism
4. a
5. c
6. protein
7. hydrolysis, digestion
8. equal to
9. b
10. c
11. decreasing
12. b
13. b
14. coenzyme, cofactor
15. b
16. a
17. cellular respiration
18. anaerobic, glycolysis
19. b
20. aerobic
21. oxygen
22. phosphorylation
23. second or aerobic
24. c
25. metabolic pathway
26. lactic acid
27. citric acid cycle, Krebs cycle
28. c
29. a, d
30. c
31. a
32. ketone bodies
33. deamination
34. d
35. rate-limiting enzyme
36. a, b
37. DNA (deoxyribonucleic acid)
38. chromatin or nucleus, cytoplasm
39. messenger, transfer
40. c
41. messenger RNA
42. a
43. b
44. b
45. ATP
46. a
47. a, c, d

5 Mastery Test Answers

1. a
2. tissues
3. a
4. epithelial, connective, muscle, nerve
5. b
6. a ,c
7. a
8. 1. e, 2 .b, 3. f, 4. c, 5. d, 6. h, 7. c, 8. g, 9. a
9. exocrine
10. a, b, d
11. fibroblast
12. macrophage
13. c
14. a
15. less
16. a, b
17. a
18. a, b, c, d
19. muscle, bone; bone, bone; fibrous
20. c
21. bone
22. plasma
23. smooth, skeletal, cardiac
24. nervous
25. neuroglial

6 Mastery Test Answers

1. serous, serous
2. mucous, mucus
3. loose, adipose, synovial
4. prevents harmful substance from entering, retard water loss, regulate temperature, house sensory receptors, contain immune cells, synthsize chemicals, excrete wastes
5. epidermis
6. dermis
7. subcutaneous layer
8. c
9. a
10. contact dermatitis
11. b
12. a
13. sunlight
14. epithelial
15. b
16. contains
17. c
18. hair, sebaceous glands, nails, sweat glands
19. dead epidermal cells
20. a
21. c
22. lunula
23. a, c

24. b
25. apocrine
26. c
27. a
28. albino
29. c
30. c
31. b
32. granulations
33. a

7 Mastery Test Answers

1. a
2. a, b, c, d
3. long
4. c
5. b
6. b
7. b, d
8. compact
9. spongy (cancellous)
10. c
11. collagen, inorganic salts
12. sickle-cell
13. intramembranous bones
14. endochondral bones
15. d
16. epiphyseal disk
17. b
18. a, d
19. a
20. a
21. b
22. e
23. strengthen or thicken; weaken or get thinner
24. c
25. a, c
26. simple
27. b
28. levers
29. a
30. bone, mineral
31. estrogen
32. 206
33. a
34. skull, hyoid, vertebral column, thoracic cage
35. pectoral girdle, arms or upper limbs, pelvic girdle, legs or lower limbs
36. b
37. a, d
38. occipital
39. aphenoid
40. maxillary
41. palatine processes
42. fontanels
43. zygomatic, lacrimal
44. c
45. b
46. spondylolysis
47. a
48. sacrum
49. c
50. a, b, c, d
51. thoracic vertebrae, sternum
52. d
53. angle of Louis or sternal angle
54. clavicles, scapulae
55. b
56. a
57. metacarpal
58. a
59. c
60. b
61. calcaneus

8 Mastery Test Answers

1. d
2. d
3. immovable, slightly moveable, freely movable; fibrous, cartilaginous
4. b, c
5. fibrous
6. synchondrosis, cartilaginous joint
7. is
8. more
9. c
10. subchondral plate
11. a
12. d
13. b
14. synovial membrane
15. a, c
16. menisci
17. bursa
18. synovial fluid
19. d
20. 1. c, 2. e, 3. b, 4. a
21. humerus and scapula
22. is not
23. dislocation
24. humerus, ulna
25. pronation and supination of the hand
26. arthroscope
27. acetabulum, coxal
28. b
29. extension, flexion, abduction, adduction, rotation, and circumduction
30. knee
31. a
32. sprain

9 Mastery Test Answers

1. a
2. b
3. fasciculi or fascicles
4. perimysium
5. a
6. fasciotomy
7. actin, myosin
8. b
9. a
10. strain
11. c
12. d
13. contraction or shortening
14. b
15. fewer
16. myosin
17. familial hypertrophic
18. a
19. dystrophin
20. stimulate
21. calcium
22. tropomyosin-troponin
23. b
24. acetylcholinesterase
25. *Clostridium botulinum*

26. adenosine triphosphate or ATP
27. creatine phosphate
28. a, c, d
29. lactic acid
30. rigor mortis
31. b, c
32. heat
33. threshold stimulus
34. c
35. myogram
36. c
37. a
38. c
39. more slowly
40. multiunit, visceral
41. a, c
42. a
43. yes
44. calcium channel blockers
45. origin, insertion
46. antagonists
47. d
48. temporomandibular joint
49. a
50. b
51. d
52. linea alba
53. b
54. calcaneal or Achilles

10 Mastery Test Answers

1. a
2. a, b, c
3. neurons and neuroglia
4. synapses
5 central, peripheral
6. sensory
7. neuron
8. soma, perikaryon
9. b
10. d
11. yes
12. a
13. c
14. c
15. a
16. c
17. c
18. c
19. b
20. c
21. b
22. neuroma
23. a
24. resting potential
25. summation
26. c
27. absolute refractory
28. b
29. a
30. a
31. b
32. b
33. a
34. c
35. enkephalins
36. facilitated
37. convergence
38. a
39. d
40. sensory, motor, interneurons

11 Mastery Test Answers

1. brain, spinal cord
2. a
3. subdural hematoma
4. c
5. b, c
6. c
7. 31
8. b, d
9. b
10. b
11. b
12. a
13. a
14. d
15. a
16. b
17. a
18. cerebrum, cerebellum, brainstem
19. spinabifida
20. a
21. c
22. b, c
23. 1. c, 2. d, 3. c, 4. a, 5. b, 6. a
24. speak
25. frontal
26. b, c
27. left
28. b
29. transient ischemic attack
30. b
31. choroid plexuses
32. b
33. d
34. hydrocephalus
35. d
36. d
37. b
38. limbic system
39. c
40. medulla oblongata
41. reticular formation
42. paradoxical
43. b, d
44. a
45. somatic, autonomic
46. c
47. 12, brain stem
48. b
49. c
50. brachial
51. b
52. autonomic
53. thorax and sacrum
54. cranial and sacral
55. a, b, a, b
56. a, b
57. c, d
58. brain, spinal cord, ganglia

12 Mastery Test Answers

1. b
2. a, b, c, d, e
3. c
4. somatic
5. 1. b, 2. d, 3. b, 4. b, 5. c, 6. a
6. b

7. b
8. d
9. c
10. d
11. a, c
12. referred
13. acute
14. spinal
15. endorphins
16. special
17. b
18. c
19. olfactory nerve or olfactory tracts
20. synesthesia
21. b
22. b
23. sweet, salty, sour, bitter
24. smell
25. equilibrium
26. cerumen
27. c, d
28. d
29. c
30. c
31. osseous labyrinth, membranous labyrinth
32. perilymph
33. a
34. c
35. conduction or conductive
36. sensorineural
37. sensorineural
38. vestibule
39. b
40. c
41. a
42. cornea
43. a
44. d
45. b
46. optic nerve
47. cornea
48. b, c
49. a
50. cataract
51. cornea
52. b
53. pupil
54. melanin
55. retina
56. b
57. vitreous humor
58. refraction
59. presbyopia
60. rods, cones
61. 1. a, 2. b, 3. a, 4. b
62. rhodopsin, opsin, retinene
63. erythrolabe, chlorolabe, cyanolabe
64. color blindness
65. partial, both eyes

13 Mastery Test Answers

1. diabetes mellitus
2. exocrine
3. endocrine
4. paracrine
5. autocrine
6. b
7. d
8. a, b, c, d, e
9. c
10. c
11. a
12. b
13. d
14. arachidonic acid (fatty acids)
15. steriod hormones
16. b
17. b, d
18. hypothalamus
19. a, b
20. posterior
21. a, c
22. b, d
23. acromegaly
24. prolactin
25. a, d
26. a, b, d
27. c, d
28. b
29. Thyroxine, triodothyronine
30. c, d
31. iodine
32. calcitonin
33. c
34. c
35. d
36. epinepherine, norephinpherine
37. a
38. a, b, c
39. male
40. b, c
41. Islets of Langerhans
42. glucogon
43. b,d
44. diabetes mellitus
45. b
46. c, d
47. a
48. Tlymphocutes
49. more
50. erythopoietin

14 Mastery Test Answers

1 b
2. hematocrit
3. plasma
4. 55
5. a
6. a, b, c
7. cyanosis
8. nucleus
9. c
10. b
11. a
12. b
13. d
14. erythropoietin
15. yes
16. b
17. porphoria
18. a
19. b
20. 5,000 to 10,000
21. b, c
22. yes
23. platelet or thrombocyte
24. 1. c, 2. a, 3. b, 4. b, 5. c

25. a
26. low-density lipoproteins
27. b, c, d
28. b
29. b, d
30. b
31. b
32. fibrinogen, fibrin
33. a, c
34. d
35. positive
36. disseminated intravascular coagulation
37. prothrombin time, partial prothrombin time
38. b
39. streptokinase
40. d
41. a
42. hemophilia
43. b
44. a
45. c
46. erythroblastosis fetalis
47. Rhogam
48. a

15 Mastery Test Answers

1. d
2. mediastinum
3. d
4. a
5. d
6. c
7. atria, ventricles
8. stretching
9. a, d
10. c
11. a, b
12. c
13. aortic
14. a
15. cardiac skeleton
16. coronary arteries
17. b
18. myocardial infarction
19. amyloid
20. decreases
21. cardiac cycle
22. d
23. b
24. a
25. c
26. decrease
27. b, c
28. a
29. d
30. b
31. b
32. c
33. a
34. endothelium
35. b
36. impermeability
37. filtration, osmosis, diffusion
38. a
39. c
40. a
41. diffusion
42. c
43. b
44. valves
45. b
46. varicose veins
47. b
48. pulse
49. stroke volume
50. heart action, blood volume, resistance to flow, viscosity, peripheral resistance.
51. stroke volume, heart rate
52. c
53. parasympathetic
54. b
55. secondary
56. a, b
57. d
58. b
59. b
60. d
61. left atrium
62. c
63. circle of Willis
64. hepatic portal system
65. a

16 Mastery Test Answers

1. a
2. lymphatic capillaries; collecting ducts
3. lacteals, fats
4. b
5. b
6. c
7. c
8. a, c
9. veins
10. a
11. edema
12. c
13. d
14. lymphadenitis
15. Peyer s patches
16. b
17. b
18. b
19. b, c
20. spleen
21. b, d
22. pathogens
23. b
24. c
25. a, d
26. c
27. redness, swelling, heat, pain
28. b
29. neutrophils, monocytes
30. c
31. reticuloendothelial
32. immunity
33. thymus gland
34. antigens
35. d
36. d
37. cell-mediated
38. gamma globulin
39. b
40. c
41. T-cells
42. d
43. passive
44. B-cells

45. c
46. b
47. tissue rejection
48. infection
49. autoimmune

17 Mastery Test Answers

1. d
2. digestion
3. alimentary canal
4. accessory organs
5. b
6. mixing, propelling
7. no
8. b, c
9. a, c, d
10. frenulum
11. c
12. wisdom teeth
13. a
14. c
15. d
16. increase
17. b
18. b
19. peristalsis
20. heartburn
21. b
22. hypertrophic pyloric stenosis
23. c
24. b
25. vitamin B_{12}
26. decreases
27. b, c
28. inhibits
29. chyme
30. a
31. inhibits
32. a
33. a
34. b
35. alkaline
36. c
37. upper right
38. c
39. ferritin
40. c
41. b, c, d
42. a, d
43. c
44. a, b, d
45. duodenum, jejunum, ileum
46. b
47. c
48. d
49. c
50. peristaltic rush
51. cecum
52. b
53. electrolytes, water
54. a, c

18 Mastery Test Answers

1. anorexia nervosa
2. carbohydrates, proteins, fats
3. vitamins, minerals
4. essential nutrients
5. b
6. a, b
7. cellulose
8. oxidation
9. b, d
10. d
11. b
12. liver
13. linoleic acid
14. a
15. d
16. cholesterol
17. too much
18. a, c, d
19. amino acids
20. complete
21. yes
22. a
23. a, c
24. calories
25. basal metabolic rate
26. a, c
27. a, b
28. negative
29. a
30. fat
31. b
32. a
33. sunlight or ultraviolet rays
34. c
35. b, d
36. b
37. a
38. c
39. d
40. c
41. calcium, phosphorus
42. c
43. b
44. aldosterone
45. sodium
46. d
47. oxygen
48. a, b, c, d
49. c
50. primary
51. b
52. b

19 Mastery Test Answers

1. a
2. c
3. energy
4. b, d
5. 1. a, 2. b, 3. c, 4. a, 5. c
6. a, b, c, d
7. no
8. nasal cavity, larynx
9. b, c
10. thyroid cartilage
11. a
12. a
13. tracheostomy
14. c
15. alveolar ducts
16. decreases
17. fiberoptic bronchoscope
18. a
19. b
20. hilus
21. right
22. visceral pleura
23. parietal pleura

24. contracts, increasing, decreasing
25. c
26. a
27. pleura
28. surfactant
29. c
30. compliance
31. b
32. spirometer
33. b, d
34. respiratory capacities
35. trachea, bronchi, bronchioles
36. equal
37. alveolar ventilation
38. nonrespiratory air movements
39. b
40. c
41. brain stem
42. c
43. a
44. carbon dioxide
45. c
46. a, c, d
47. alveolar macrophage
48. alveolus, capillary
49. pressure
50. partial pressure
51. a
52. hemoglobin
53. increase
54. hemoglobin
55. c

20 Mastery Test Answers

1. d
2. b, c
3. b, c
4. renal pelvis
5. d
6. Wilms
7. b, c, d
8. d
9. a, b
10. b
11. b
12. d
13. urine
14. a
15. b
16. hydrostatic pressure
17. decrease
18. decrease
19. decrease
20. yes
21. c
22. d
23. b
24. a
25. b
26. aldosterone
27. countercurrent
28. b
29. protein
30. d
31. b, d
32. d
33. d
34. b
35. ureters
36. c
37. decrease
38. trigone
39. b
40. urgency
41. 150cc
42. a
43. vagina, clitoris

21 Mastery Test Answers

1. c
2. equal
3. c
4. intracellular
5. extracellular
6. d
7. hydrostatic
8. osmotic
9. a
10. c
11. c
12. diabetes insipidus
13. extracellular fluid
14. c, d
15. enters
16. a
17. decreased; osmotic
18. food, beverages
19. perspiration, feces, urine
20. aldosterone
21. a, b, c
22. a, c
23. hydrogen ion
24. c
25. 7.4
26. more
27. b
28. c
29. combine with
30. c
31. a
32. rate, depth
33. c
34. ammonium
35. a
36. b

22 Mastery Test Answers

1. d
2. b
3. c
4. a
5. testes
6. b
7. b
8. spermatic cord
9. a
10. spermatogonia
11. four
12. 23
13. continuously
14. c
15. prostate gland
16. b, d
17. emission
18. b
19. c
20. a, d
21. dartos
22. b

23. testes, hypothalamus, anterior pituitary gland
24. c
25. secondary sexual
26. inhibitin
27. ovaries
28. d
29. one
30. primary follicle
31. b
32. d
33. ovulation
34. a, d
35. endometrium
36. a
37. b, c
38. c
39. a, c
40. more
41. estrogen, progesterone
42. b, c
43. a
44. b
45. corpus luteum
46. b
47. a, d
48. c, d
49. menarche, menopause
50. d
51. c
52. one
53. b
54. placenta
55. b
56. progesterone, oxytocin
57. c
58. prolactin
59. a
60. tamoxifen
61. c
62. a
63. b, c
64. b
65. vasectomy
66. c
67. c

23 Mastery Test Answers

1. growth
2. development
3. prenatal
4. postnatal
5. c
6. a
7. a
8. b, d
9. in vitro
10. primary germ
11. b
12. b
13. a
14. placenta
15. amnion
16. c
17. d
18. eight
19. teratogens
20. neural tube defect
21. b
22. b
23. b, d
24. vein
25. a
26. foramen ovale, ductus arteriosus
27. surfactant
28. b
29. d
30. less
31. d
32. infancy
33. adolescence
34. senescence
35. a
36. b

24 Mastery Test Answers

1. human genome
2. heredity
3. environment
4. c
5. human genome
6. 23
7. c
8. b
9. d
10. c
11. b
12. incompletely dominant
13. pleiotropy
14. genetic heterogeneity
15. c
16. c
17. d
18. a
19. wild type
20. c
21. polygenic
22. c
23. sex linked
24. a
25. trisomy
26. medical genetics
27. b, c
28. amniocentesis; chorionic villi biopsy
29. heritable

CREDITS

p. 112: From Kent M. Van De Graaff, *Human Anatomy,* 3d ed. Copyright 1992 Wm. C. Brown Communications, Inc. Dubuque, Iowa. Reprinted by permission of Times Mirror Higher Education Group, Inc., Dubuque, Iowa. All Rights Reserved.

pp. 162, 163: From Kent M. Van De Graaff, *Human Anatomy,* 3d ed. Copyright 1992 Wm. C. Brown Communications, Inc. Dubuque, Iowa. Reprinted by permission of Times Mirror Higher Education Group, Inc., Dubuque, Iowa. All Rights Reserved.

pp. 186, 188: From Kent M. Van De Graaff, *Human Anatomy,* 3d ed. Copyright 1992 Wm. C. Brown Communications, Inc. Dubuque, Iowa. Reprinted by permission of Times Mirror Higher Education Group, Inc., Dubuque, Iowa. All Rights Reserved.

pp. 215, 216: From Kent M. Van De Graaff, *Human Anatomy,* 3d ed. Copyright 1992 Wm. C. Brown Communications, Inc. Dubuque, Iowa. Reprinted by permission of Times Mirror Higher Education Group, Inc., Dubuque, Iowa. All Rights Reserved.

p. 247: From Kent M. Van De Graaff and Stuart Ira Fox, *Concepts of Human Anatomy and Physiology,* 3d ed. Copyright 1992 Wm. C. Brown Communications, Inc. Dubuque, Iowa. Reprinted by permission of Times Mirror Higher Education Group, Inc., Dubuque, Iowa. All Rights Reserved.